T0304556

FRANCESCO VERSO

THE ROAMERS

Translated by Jennifer Delare

This is a **FLAME TREE PRESS** book

Text copyright © 2023 Francesco Verso
Translation copyright © 2023 Jennifer Delare

FLAME TREE PRESS
6 Melbray Mews, London, SW6 3NS, UK
flametreepress.com

US sales, distribution and warehouse:
Simon & Schuster
simonandschuster.biz

UK distribution and warehouse:
Hachette UK Distribution
hukdcustomerservice@hachette.co.uk

Publisher's Note: This is a work of fiction. Names, characters, places, and
incidents are a product of the author's imagination. Locales and public names
are sometimes used for atmospheric purposes. Any resemblance to actual
people, living or dead, or to businesses, companies, events, institutions, or
locales is completely coincidental.

Thanks to the Flame Tree Press team.

The cover is created by Flame Tree Studio with
thanks to Nik Keevil and Shutterstock.com.
The font families used are Avenir and Bembo.

Flame Tree Press is an imprint of Flame Tree Publishing Ltd
flametreepublishing.com

A copy of the CIP data for this book is available from the British Library
and the Library of Congress.

1 3 5 7 9 8 6 4 2

HB ISBN: 978-1-78758-834-9
PB ISBN: 978-1-78758-833-2
ebook ISBN: 978-1-78758-835-6

Printed and bound in Great Britain by Clays Ltd, Elcograf S.p.A.

FRANCESCO VERSO

THE ROAMERS

Translated by Jennifer Delare

FLAME TREE PRESS
London & New York

PART ONE

THE PULLDOGS

'This life, which is such a fine thing, is not the life we are
acquainted with, but that of which we know nothing; it is
not the past life, but the future.'
G. Leopardi, 'The Dialogue Between an Almanac Seller and a Passer-By'
in *Moral Essays*, 1834 (from the translation by Charles Edwardes,
Edwin and Robert Grabhorn, San Francisco, 1921)

'There is no longer any need to talk about that which exists.'
Le Monde, September 19, 1987

Humanity spent centuries seeking out sunlight, sunlight that had been
trapped inside the Earth for millions of years in the form of vast quantities
of coal, oil and natural gas. This seemingly unlimited resource was first
used in the steam engine, then in the dynamo, and finally in the internal-
combustion engine, ushering in what was known as 'material progress'.

However, to better continue to exploit these sources of energy,
millions of human beings were uprooted from their lands and homes
and forced to resettle elsewhere in the hope of making a better living.
That illusion crumbled in the face of poorly lit factories and cramped
offices, far removed from the changing seasons and the ancient customs
of rural civilization.

This period, called the Industrial Age and recognized as the final stage of the Fossil Era, lasted for five centuries. It spread over every continent, radically altering human lifestyles as well as the planet's homeostatic equilibrium. However, during the final decades of this historical period, a new and revolutionary approach to human labor and diet was born. It arose from a series of phenomena that went far beyond any prediction, which were destined to alter mankind's very relationship with the Earth. At the beginning of the third millennium, two events occurred which, in acknowledgement of their singularity, came to be known as the First and Second Logical Mutations. These Two Logical Mutations were followed by the period in which we are now living, the Drift.

The causes of these phenomena, which gradually expanded beyond their local origins, lie in the dissemination of the first Public Matter Compositors (PMCs) and the home- and portable-nanomats that soon followed but, most importantly, we can trace their beginnings to the invention of *nems* (nano-electromechanical systems), also known as *nanites*.

The first symptom of radical change manifested as a shift in attitude towards paid employment. Work had stopped being capable of helping man to achieve their desires. It served to satisfy either exclusively those needs which occupy the tier that is the first and lowest of the hierarchy illustrated in Maslow's pyramid (i.e. the physiological needs: breathing, food, sex and sleep) or extended to include the second (i.e. safety needs: security of body, of employment, of the family, of health and of property). According to statistical indices dating from that time on numbers of marriages and divorces, social volatility and family resilience, only a tiny percentage of the global population could feel assured it would achieve the third tier (i.e. friendship, family relationships and sexual intimacy). Cases of people who could say they had been able to attain the fourth tier (self-esteem, self-control, achievement, respect of oneself and others) and remain there for more than a few years were rare indeed.

The fifth level (morality, creativity, spontaneity, problem solving, acceptance, lack of prejudice) was not believed to be achievable through work. In fact, anyone who enjoyed that rarest of conditions had certainly arrived there by other routes, the first among these being the opportunity to give up work as it had always been understood – or perhaps misunderstood.

The second change involved human diet. For thousands of years, food had remained a reflection of the societies that harvested, produced and consumed it. It provided the substance and the ideas needed for the evolution of every known civilization, but it also supplied the unique mechanisms that allowed for their eventual divorce from it and resulting transformation.

Many of the things people thought of as pleasurable – family relationships, cultural identity, ethnic diversity – were all intimately linked to food preparation and consumption. However, these had begun to undergo a rapid transformation as a result of habits such as dining out in restaurants or eating in company cafeterias. Not only had people begun to cook less, but the number of people who knew how to cook at all was growing steadily smaller. The 'outsourcing' of the food preparation process to manufacturers and multinational companies like Nestlé, Kraft, Unilever and Dannon was the first step towards the disappearance of culinary traditions. The Internet could provide no help in reversing that trend. The recipes it offered were based on abstract instructions, citing ingredients hardly anyone was familiar with anymore and techniques that had fallen out of use among large swathes of the population. To whisk, to nap, to let meat become high, to brown, to parboil – this lexicon lost its meaning once the manual skills associated with it had vanished. What's more, the wait (meaning the time needed to cook a given food) was a foreign concept in that modern culture. Botched attempts ended up in the trash, replaced by a product from the fridge or a local delivery restaurant.

The number of people who saw food preparation and consumption as a pastime or a means to socialize and amuse themselves continued to shrink. At that point, food preparation time ranged from five to fifteen minutes, while the number of meals not cooked in a microwave or composited by a nanomat grew progressively smaller.

By 2019, the most common 'meal' in the world was the sandwich. Meals became snacks, restaurants shrank into snack-bars, cafeterias were replaced by vending machines, and offices were furnished with 3D food printers. When the first 'squeeze-packs' arrived, complete meals you could carry in your pocket and squirt down your throat, it became clear that, in the future, the act of eating would be purely incidental.

When it had ceased to serve its original purpose, the act of human reproduction had been transformed into a game, a recreational activity. Human nutrition underwent the opposite evolution, in that people no longer ate for the pleasure of it, but only to resupply themselves with nutrients and energy.

Beginning in the third millennium, food gradually and inexorably began to vanish. The first signs of this date back to the late nineteen- hundreds, when multi-course meals began to be superseded by the single course, fast food by microwave cooking, fun food the single-serving meals served on trains and airplanes, delightful gifts/diversions whose enjoyment lay chiefly in discovering what was inside them as opposed to in eating them – by the finger food typical of happy-hour buffets and caterers' trays. The end results of this process were rapidly metabolized *nutraceuticals* and foods composited by 3D printers. Paradoxically, this transformation went hand in hand with the phenomena of obesity and malnutrition, which were a continual torment for over half the world's population.

This is the story of a group of people, the Walkers, but it is also the story of an anthropological transformation that would forever alter civilization as we know it, giving rise to a new – though in some ways ancient – form of culture, a nomadic and creative society that revolved around the sun.

PHASE ONE
THE FIRST LOGICAL MUTATION

'Too much speed is like too much light…we see nothing.'
Paul Virilio

'If you have built castles in the air, your work need not be lost; that is
where they should be. Now put the foundations under them.'
Henry David Thoreau, *Walden*

MIRIAM FARCHI

CHAPTER ONE

Globalzon

In the month of March the skies of Rome offer up an incredible show, free of charge. Not the usual haze of smog, but thousands of birds – swallows, sparrows and starlings – gather above the city's roofs and umbrella pines for a respite from their journey, a stopover before continuing on their migration.

Their swooping encounters trace fantastical choreographies in the sky, acrobatic evolutions, mosaics whose patterns are ever changing, aerial itineraries of astounding fluidity and vitality. Suddenly they veer, breaking cleanly into two or three neat flocks. They follow the updrafts, each knowing exactly when to glide and when to climb only to once more join and blend into the whole.

Miriam Farchi's face is tilted skywards. Each time she finds herself tracking those perfect movements – coordinated by goodness knows what instinct, order or rule, without any sort of leader or flight director to issue instructions – she forgets to chew.

One day, she had been watching a swallow, a specimen with a ring of red feathers around its throat. She had seen it swoop in to take the head of the flight and, for a short while thereafter, guide its wheeling. It had been a commonplace scene, one doubtless repeated untold times, yet somehow both extraordinary and moving. At that moment, Miriam had come to the realization that any group as dense and compact as that of the birds functions because each one of its components, whether at the center of the formation or at its fringes, takes a turn at making the decisions. Later, she had done some research and she had learned that no one knew how to explain the dynamics of those airborne acrobatics. They could be a sign of harmony within the birds' society just as easily as they could be an expression of internal strife.

Miriam finally swallows her mouthful of Cocorich – a bar enriched with nutraceuticals and composited, as the wrapper informs her, in a laboratory called U46G-PRC – and recalls the passage of time. Her lunch break is about to end. A pair of street cats circle around, taking turns rubbing up against her legs to collect the crumbs from her snack-bar.

Miriam opens her purse, takes out her smartphone and switches it off silent mode. Nearby, on the other stone benches that line Piazza Bernini all'Aventino, some of her fellow World Food Programme colleagues are doing the same. Four missed calls in a quarter of an hour, the last from two minutes ago. Such persistence would suggest that it's something important.

When she reads Alan's name on the display, she starts to worry. Her son has often told her that there's no point in calling him at work, since it's forbidden to take smartphones into the warehouse. All personal items have to be left in external lockers, on pain of receiving a reprimand. He has told her, resigned, that the scanners at the entrance to Globalzon ensure compliance with that rule. The ones at the exit, meanwhile, check that no one makes off with a game console for their children or a pair of sneakers or some undergarments for themselves.

It is with apprehension that Miriam dials Alan's number. She's alone now, the only one still lingering in the square.

"Hello?"

The voice she hears on the other end of the line does not belong to her son. "Who is this? Where's Alan? This is his mother."

"Oh, it's you, Mrs. Farchi. You see…my name's Giulio. I'm a colleague of Alan's."

"Hello, Giulio. Why are you answering Alan's phone?"

"Alan told me to call you, but when I tried before you didn't answer."

Instinctively, Miriam gets up from the bench and starts to walk, without knowing where she's going. "What's happened? Pass me my son, please."

She makes her way across the street, blindly. A rickshaw driver – one of many who have added their services to Rome's range of transportation offerings in recent years – launches a curse in her direction. The passenger, a businessman with an air of confusion, most likely an executive visiting from abroad, glances at the time, looking annoyed.

"You see, Alan's had an accident."

Miriam falters. She sits back down, on the bench nearest to hand. "What sort of an accident?"

"It was an electrostatic discharge. I've been hit by quite a few myself."

"So it's not serious, then. Is he all right?"

"I don't know. He was up on level D, twenty feet above the floor. I was behind him and saw him lose his balance. He tried to find something to grab on to but then…he fell."

Alan has always hated that job. Every time they see each other, every time he complains, Miriam tries to persuade him that working – even in a distribution warehouse like the one owned by Globalzon – is better than sitting at home, shaking the remote control and cursing the economic crisis while wasting away. Not even playing the guitar, his greatest passion, is a comfort to him any longer.

"How did he fall? Did he get hurt?"

"That's the problem. He hit the arms of the freight hoist going down, the one we use for lifting pallets. He hit his back."

Miriam squeezes her eyes shut in an effort to stave off panic. She takes a deep breath before speaking. "Where is he now? Is he at the hospital? Can I talk to him?"

"No, he's here, at the warehouse. When I went to help him, he told me he doesn't have any insurance, so he can't be transported. He told me to call you. Hang on, I'll pass him the phone."

Miriam listens, trying to make out what's happening. In the background she can hear the sounds of tracked vehicles and irregular beeping noises.

Alan has only been working at Globalzon for a few months. Its Roman hub lies inside one of Tiber's loops, near the Marconi Bridge. It's a shipping company but, unlike its competitors, it has invested next to nothing in machinery, opting for more inexpensive human labor. Well, at least that's what Alan told her, when he first accepted an on-call contract to help cover the Christmas-holiday peak times.

"Mom…."

"Alan, how are you? What happened?"

"A disaster." His voice is fractured, pained. "Listen, these assholes say they won't call me an ambulance. They say…that they can't let anyone into the warehouse."

"But, how are you? Can you stand?"

"No, my legs…I can't feel them."

The fear of what may have happened renders Miriam speechless.

"Call someone, Mom. Hurry! They say that the most they can do is… leave me outside the gate. That's already doing me a favor. They can't stop working."

Miriam gives herself a shake, gets up and starts to run towards Viale Aventino. There's a cab stand near her office.

"What's the exact address?"

"Lungotevere Dante, number 34."

The immense flock of birds has disappeared. Maybe the presence of some predator has forced them into an emergency landing. Only two pairs still linger in the sky to the south, somewhere above Ostiense Station.

"I'll be there as soon as I can."

Miriam ends the call then pulls up the contact information for Cecile, her supervisor in the WFP's food analysis division. A brief explanation why she's late coming back from lunch earns Miriam her boss's understanding in the form of a half-day off. When she sees the traffic lined up at the info-signal in Piazza Albania, she changes her plan and types into Google, 'fast transport Rome'. Up pop ads for the Speedy Boys, Bartolini and UPS. Beneath those is an ad for the Pulldogs. Miriam opts for the least orthodox but most efficient solution.

A rickshaw picks her up in front of the Piramide subway station three minutes later. Miriam is shaken, sweaty from running. As soon as she recognizes the Pulldogs' symbol – a stylized dog with a tow harness fitted to a bit in its mouth – she climbs aboard.

"And the other one? The one for my son?"

"It's waiting for us, further along, on Via Ostiense. From there we'll go on together."

"Please hurry. My son has had an accident."

"You told us that, ma'am. Now hold on tight."

The girl behind the push bar of the rickshaw – a ramshackle contraption the color of dirty silver – is tall and brawny. There is something masculine about her toned and well-defined bands of muscle, visible beneath skintight cycling clothes. On her head she wears a white bandanna with a floral pattern. An anti-smog mask covers the lower half of her face. With a shove against the push bar, she surges out into the traffic, or rather, into one of the open passageways between cars.

Miriam clings to the central handrails – two barbells without their weights, wrapped in red leather. "How long till we get there?"

"It depends, but no more than five minutes."

The other rickshaw, a streamlined model with chrome wheels and a seat designed for a race car, is waiting for them like a relay runner. It's already begun to move when Miriam picks out its canopy from a distance. A boy is pushing this one, his face pitted by teenage acne. He's so skinny that it's hard to tell where he gets the strength to haul his rickshaw from morning to night. Still, both drivers keep on, agile and quick, never stopping.

Ahead, along a loop in the Tiber where the old dog-racing track once stood, Miriam can make out the nondescript outline of Globalzon, transit point for every sort of product under the sun. Once it had absorbed the manufacturing sector's supply chain, it had gone on to swallow up the food industry as well. Outside there are no signs or indications to distinguish it from any other warehouse. Alan says that's for security, to protect the company and its employees. There have been reprisals and revenge attacks, after all. Every so often, management drops by for an unannounced inspection. Ignorance maintains apprehension and apprehension keeps people on their toes.

Giulio is kneeling in front of the bar that blocks the entrance, keeping watch over Alan, who is lying on the ground by the edge of the street. On high, the seagulls cry and wheel above the warehouse. A number of forklifts stand parked in the service area, while stacks of pallets are being unloaded from a pair of tractor trailers.

When he sees Miriam, Alan finally stops trying to keep it together. He's been holding on for nearly an hour, but he just can't do it anymore. He smiles at her, a bit of happiness clinging to his lips, then he faints. She wishes she could to talk to him, check how badly he's hurt, but that hope is denied.

"Where do we take him? San Camillo?"

The girl's voice sounds distorted, as though it's coming through a dead megaphone. Everything has been happening so fast that Miriam hasn't even stopped to think about which hospital is closest.

"Yes, you're right. San Camillo."

Miriam is about to lift her son, but the two Pulldogs get there first.

"Let us do it. Come on, Little Simon. You drive the lady and I'll take him. He weighs more."

Together, they lift Alan and ease him down into the first rickshaw. His feet roll back and forth, then lie still. Miriam has to resist the temptation to pull up his shirt and look for something, a cut or an injury of some kind on his back. Instead, she thanks Giulio and climbs into the other rickshaw, her foot already on the running board as she waves.

The girl takes the bandanna from her head and knots it around the antenna like a white flag. Then she's off, lunging forward.

"What happened to him?" she shouts, pulling alongside Miriam's rickshaw as they slow down at the entrance to the Marconi Bridge.

She must be the same age as Alan. She's covered with tattoos and, now that nothing is covering her head, Miriam can see that its sides are shaved, leaving only a thick black strip of hair, a mohawk-style crest that gives her a fierce look, well suited to her profession.

"I don't know for sure – just that he fell and he hit his back. He can't feel his legs."

The girl looks down at her own legs, two powerful pistons propelling her nimbly through the traffic on Lungotevere degli Inventori – legs that dodge left and right, brake and bend. Every once in a while they jump up onto the sidewalk, swerve suddenly, veer into the opposite lane going in the wrong direction. The legs of the second driver, behind them, do the same.

"I've heard a lot of rumors about that place."

Miriam doesn't know how to answer. She jolts up and down and holds tightly to the handgrips.

"From what I hear, it's run like a military base. They give you objectives every hour, and if you achieve them, they either change them or increase them. Plus, they hire the fewest human beings they can to fill the day's orders. When they're done with you, they fire you."

Miriam isn't capable of adding anything to the conversation. She's almost not listening, but her silence seems an expression of assent.

"What bastards. If it were up to them, they'd just leave a person to die."

A few hours later, the rickshaw girl is still waiting outside the San Camillo hospital. She's alone there. She's been passing the time drinking red wine from a plastic bottle without a label.

"How's your son?"

Miriam's expression leaves no room for illusions. "Not well, but he's alive. The diagnosis will take a few days. They need to give him a CAT scan and an MRI. That's what they told me."

"I'm sorry. Listen, ma'am, hop aboard and I'll give you a ride home. Where do you live?"

"Thank you. You're very kind. I live on Via Satrico, near Porta Metronia."

It is only now, when her stomach growls for the umpteenth time, that Miriam realizes she hasn't eaten since lunchtime. As soon as she's sitting in the rickshaw, she opens her purse and takes out a Cocorich bar. Meanwhile, the girl is plowing through the traffic on Circonvallazione Gianicolense, a badly maintained ring of asphalt populated by resigned vehicles. Their drivers, however, are agitated, a herd of furious commuters who refuse to be resigned, cursing, laying on their horns and fuming. With pointing fingers and wide-open mouths, people urge each other forward with gestures and hostile glares. The more civil drivers use music to numb themselves or else lost in their smartphones, shrewdly avoiding any sort of exchange of emotion with their fellow motorists.

The rules of the road apply only to human beings. Cars, left to the guidance of apps installed in their dashboards, the first Traffic Intelligence Agents, have no need of them.

At the first red info-signal, the girl turns around and notices the wrapper left in the trash container. "You like that stuff?"

Miriam stops chewing. She doesn't really *eat* those bars. She just uses them to keep her stomach from growling and to give her enough energy to get over the hump between lunch and dinner.

"I'm used to them. I've been getting them for years. This brand is my favorite. They fill you up because they contain more protein, which gets time-released into your stomach at a rate of four calories per minute. The others are all full of sugar and after an hour you're hungry again."

"Ah, you're an expert."

"I work for the World Food Program, analyzing foods."

"What can you tell me about fats?"

"They're treacherous, because they have a rate of two calories per minute, but the signal to tell you you've had enough is slow to arrive, so you can keep eating without feeling full. It's a conundrum."

"I'm sure you're right, but I'm still against nutraceuticals. You know what really gets me? My mother runs a restaurant in Trastevere, called Il Romoletto. It's on Via della Lungara. Have you heard of it?"

"Il Romoletto? Yes, I know it. I've been there with my coworkers. Roman cuisine."

"Yeah, right. *All farm-fresh ingredients.*"

Her tone is mocking. Miriam would like to ask her why she doesn't work at the restaurant instead of driving a rickshaw, but judging from her attitude and appearance, it's likely there's some sort of conflict surrounding the issue. Whatever the case may be, the girl is clearly feeling talkative.

"It's a shame that I can't stand that stuff, either."

"Why not? The ingredients are all natural."

"Because my mother and father – he ran the place before she took over – have never really cared about quality. Sure, the ingredients come from the country. They're even 'organic', to hear her tell it, but that's only because that's how you make money off stupid tourists. If it were up to her, she'd just order mass quantities from China or make everything up herself with one of those 3D printers. You know the ones I mean?"

"So what do you eat? You must get your nutrition somehow."

"Me? Well, if the food industry had to depend on me, it wouldn't last very long. People who buy food at the supermarket eat to keep the industry going, not the other way around. I eat—"

The girl hauls on the brake, leaps forward and braces her feet against the ground.

Miriam has to grab onto her shoulders to stop herself from flying out of the rickshaw.

"Fuck you, you piece of shit!"

Just past Ponte Sublicio Bridge a car has cut right in front of them without signaling. It's a sports car, a BMW, with wheels that extend beyond the chassis and a spoiler that lights up when the driver brakes. Like every other motor vehicle, it's wrapped in a thin, greasy coat of hydrocarbons – you would be able to tell just by running a hand over it – and with every acceleration it releases fumes that are imperceptible to the eyes, but not to the nose.

"Sorry," the girl says, "but that, right there, is something else I hate – car drivers who think they own the road. But getting back to what we were saying, I only eat what I grow myself. Me and some friends of mine grow vegetables outside of Rome, on a little abandoned farmstead in Serra Spino where we've been staying for the last few years."

The girl doesn't seem tired after fifteen minutes of running up and down the hills of Rome. She's sweating and she's a little out of breath, but not so much that she has to stop to rest. What's more, she clearly couldn't care less about traffic signs or the warning messages flashing across the screens of the info-signals. She doesn't pay attention to stop signs, nor does she concede the right of way to anyone except for her fellow rickshaw drivers – who shoot past on both sides of the roadway every so often – and the poor pedestrians, forced to humbly prostrate themselves before all the rest. Any vehicle with a motor seems to have a black mark in her personal traffic code.

The first rickshaw service, a cooperative venture set up and run by ex-convicts, had appeared in Rome as a tourist attraction, an odd and whimsical means of moving around the city. The old horse-drawn carriages of yesteryear, a protected and anachronistic business, had not been able to compete with the new 'people porters' in terms of manageability and cost. Once upon a time, the job was viewed as humiliating and underpaid, but now, in a labor market suffering from permanent crisis and recession, it has gained dignity – a means like any other to make a living and be one's own boss. It attracts not only multitudes of immigrants, but also young Italians and students in search of part-time work.

"And you come all the way here and go back out there every day? You must really like this job."

"I do. Running around on a beautiful spring day in Rome is nothing to complain about."

Miriam thinks of Alan, of the stories he tells about *his* job, a nine- to ten-hour forced march, euphemistically called a 'shift', armed with a scanner in one hand and an order-chart downloaded from the Globalzon database in the other. If the company had equipped him with one of those pairs of enhanced-reality glasses used for locating merchandise, like in a video game, maybe nothing would have happened to him. Alan's job was to fill the greatest possible number of shipping orders – first locating the products, then collecting them, packing them, labeling them and shipping them. If you fell behind in your schedule (a program pre-calculated the time it would take to go from any one point in the warehouse to another), you received a half-penalty. Three penalties in a month and you'd be fired on the spot, automatically. The only rule to surviving longer than a week in there – to hear Alan tell it – was simple: 'Leave every crumb of pride and hint of a personal life outside.'

"Plus," the girl is saying, "if you can be outside, in the rain or the summer heat, while everyone else is hiding out inside air-conditioned malls or in their cars – the 'coffins of the 'shopping dead', we like to call them – then you must be healthy, right?"

"I guess so. But don't you get tired, being out in the traffic all day?"

"Ma'am, for us there's no such thing as traffic. On a good day, we think of it as nothing more than a sparring partner, an opponent to beat to the punch, to challenge in one lane or the other, to pass up when the info-signal's *green*."

The girl comes up alongside an enormous tractor trailer. She lets it pass her until she's even with the corner of the bumper, then grabs on to it. She slips a climber's grappling hook into the rear fender bracket, then lifts her feet up onto the rickshaw bar and pretends to relax, her fingers laced behind her head. She takes the opportunity to tilt her head back and stretch, then continues the conversation.

"You see, we rickshaw drivers are like a flash of chaos inside a sad, slow organism that drags itself along inside metal boxes monitored by GPS systems, taxed at every restricted-traffic zone or highway toll booth, oppressed by gas prices and the cost of traffic permits and mandatory inspections…and for what? A never-ending traffic jam."

Miriam can't help but smile at that description.

The girl unhooks her grapple, lowers her feet back to the asphalt and reclaims control of her rickshaw. She parts ways with the tractor trailer, which continues on along the Tiber, while she veers off onto the road that slopes up alongside the Circus Maximus. Their conversation seems to have fueled a burst of new energy.

"See, after expenses, even if you don't 3D-print your own rickshaw, you can lease one for just a few euros a day. Not even the rich cooperatives in the park at Villa Ada or on the Pincio hill charge more. That's where they live the good life, ferrying around high-class tourists who want to see the sights, drinking away their tips in the shade in Villa Borghese park. They aren't a real public transportation service, not like us."

"But can you make a living?"

"Believe it or not, after the first two or three months, this work pays better than a cab, a minivan or a Segway. And if you can assemble a rickshaw of your own, using one of those Matter Compositors that are popping up everywhere, it turns into an investment."

"But for how long do you think you'll keep on driving a rickshaw?"

"I don't know, but I do know it would be hard to show up for a job interview looking like this." She turns her head and runs a hand through her mohawk. Judging by the crinkling around her eyes, she must be smiling under her face mask. "I'm not some domesticated animal, with weaker sight and hearing than its wild cousins. As soon as they get used to the city, they lose their ability to survive elsewhere and become…*tame*. The same thing happens to workers and to consumers."

"Perhaps you're right. Alan sees things the way you do. He's always been a free spirit. He traveled a lot when he was younger, on those trains… Interrail, I think that's what they were called. When he was a little boy, he wanted to be either an explorer or a musician, because, he said, they 'go on so many tours'. Then he changed. He got cynical, and he never found any job he liked. He's been a bartender and a real estate agent. He's worked in a call center, an ice-cream parlor and a gas station, and now I doubt he'll ever…."

The girl slows down to turn onto Viale delle Terme di Caracalla. Miriam's office is there on the right – a branch office. The company headquarters are at Parco de' Medici. Alan's operation was going to change her life. She would have to move into his apartment, take care of him and hope for a remedy that did not exist. The seriousness of what had happened was going to make her as important a figure in her son's life as she had been when he was a child. Even more than when she had been helping them both to deal with her separation from Sergio.

A few minutes later, the rickshaw is approaching Porta Metronia. Doves flutter above the Aurelian Walls. This year, a colony of a hundred or so has chosen the Appio Latino neighborhood as its nesting grounds. Sometimes Miriam sees them appear as a dense cloud swirling in the sky, only to disappear as though it had never been. Other times they streak down to land in the strips of grass that run along the base of the walls, searching for seeds and shoots.

The girl stops, walks around behind the rickshaw, opens the rear luggage compartment and pulls out a pair of fine screens, rigged onto rods like butterfly nets. She goes back to pushing the rickshaw, at the same time skillfully maneuvering the screens, first along the ground, then along the edges of the sidewalks, and finally in the air, brushing them along the sides of the guano-spattered parked cars.

"You don't mind if I collect a little fertilizer for my garden, do you?"

Miriam offers a hint of a smile in response. "I see what you're doing, you know, with all of this talk about traffic, about food. It's kind of you to distract me, and here I haven't even asked you your name."

The rickshaw slows, then stops, this time in the middle of the road.

"Don't mention it. Most of the people I take around, I wouldn't want them to know the first thing about me, but you're different. Whenever you want to go back to San Camillo, call me. Either that, or put my name in the text message you send to the Pulldogs. I'm Silvia. Silvia Ruiz."

As though her words are not enough, the girl pulls off her glove, turns around and extends her hand. When Miriam takes it, she gets a vigorous handshake in return. The energy of the run is still coursing through the girl's fingers.

"My name is Miriam Farchi."

CHAPTER TWO

Ivan Shumalin

From the windows of her office on Viale Aventino, Miriam watches enormous multicolored clouds floating above the Circus Maximus, in shades enhanced by mauve, indigo and rust-colored additives. Beneath them, the wizened crowns of the Mediterranean pines sway limply to the rhythm of a tepid breeze. When she turns back towards her colleagues, she ignores their probing looks. It feels to her as though no event, either human or natural, can touch her. Since the day of the accident, she has been searching for any bit of information that could help restore to her disabled, paraplegic son the use of his legs.

Miriam takes her smartphone from her purse and types out a message. She rereads it, changes a few words, then sends it.

According to the medical report, Alan has a 'fracture of the vertebrae of the lumbar region, from L1 to L5, caused by a simultaneous bending and twisting of the spinal column resulting from an impact with a metal object'. The spinal ligaments are torn and the spinal cord is damaged. Diagnosis: 'Paraplegia of the organs below the waist, affecting the intestines, bladder and sexual organs'. It is a sentence – in every sense of the word – that Miriam has read and reread until she knows it by heart.

Around one o'clock, when the others leave for lunch, Miriam remains at her desk, surfing alternative medicine websites and the sites of clinics specializing in the treatment of spinal cord trauma.

Bones do not regenerate themselves. It seems final. That's how it is. Fractures can only mend because new bone matter, produced by other tissue, seals and shores up the broken edges of a break. While people occasionally talk about bones regrowing, human bone does not possess that ability, although many living species – including plants, amphibians and some mammals – do. In cases such as those, when an extremity, a limb, or even the spine is broken, some of the periosteal cells remain where they are, while others migrate to the blood clot surrounding the fracture, where they are transformed into osteoblasts, or bone cells. The osteoblasts form a hard ring around the fracture, repairing it, while the bone marrow produces new 'undifferentiated' tissue. The state of this new tissue is primitive, neo-embryonic. It can become connective tissue, in the form of premature cartilage, then mature cartilage and, finally, bone tissue, which can aid in the recomposition of the fractured area from within.

That is how true bone regeneration works, the kind that never occurs in humans. At best, one of two similar things occurs. The first is physiological repair, whereby small wounds or worn tissue are replaced by nearby cells of the same kind, which simply multiply to close the break. The second is scarring, in which a wound that is too large for the first method is covered over with collagen fibers.

Technology has, however, provided alternatives, and the best place to find these in Rome – or in Italy, for that matter – is the Santa Lucia Scientific Institute for Research, Hospitalization and Health Care. Doctors there have developed multiple interfaces for brain–limb communication, ones that bypass the damaged spinal cord. Their Mindwalker 3 is an eleven-pound marvel, whose parts, with the exception of the electronics, can be 3D-printed. The result is a perfectly ambulatory human being, who can walk thanks to a slow, funny looking and slightly clumsy exoskeleton.

Another solution, developed by the National Institute for Workplace Injury Insurance, consists of *replacing* a limb with a biomechanical prosthesis. In that scenario, original motor function is restored but, over time, problems may arise linked to rejection of the artificial part.

Miriam, knowing her son, is skeptical. Alan would never accept having to walk around inside an exoskeleton, not even if they could find the money to buy one. He and machines are sworn enemies, and his feelings on the matter run deep. The reason for his enmity dates back to his childhood and Miriam feels, at least in part, responsible. As a boy, Alan, like so many children, loved to play hide-and-seek. Every time someone rang the doorbell, he'd run and slip under a bed, behind a door or into a closet. He wanted to be found, and he would not answer when called, not until his hiding place was discovered. Miriam had often become angry and scolded him for being so eager to find a good spot that he wound up somewhere that could prove dangerous – up the ladder to the loft in their apartment, on a carousel in the park or even in the street. Sergio had always taken Alan's side. It was, after all, one of the few playful connections he had managed to establish with his son – this masculine game that simulated a hunt. The problem was that Alan loved to be hunted, and would force whoever happened to be there to seek him out.

On that day, she and Alan had gone to the grand opening of Com-pro, an enormous supermarket where the products were 'alive'. They talked and sang their own praises, some when you simply touched them, but others with even less prompting. In those days, grocery shopping had still been somewhat similar to the old-fashioned experience of roaming the aisles in search of products, the only difference being that *voices* and information came from the Food Intelligence Systems either concealed in the packaging or attached to it using water-soluble adhesive chips. Once you picked out your products, you walked your cart through a fast-pay food-detector. These systems, which had replaced the old cashiers, would debit the amount to be paid directly from your account.

Alan had been in the saddle of an electronic horse for over half an hour and he was tired of sitting still. When they got to the exit, he slipped free of Miriam's hand and ran away, shouting, "Catch me! Come on! Catch me!"

Miriam called to him to come back, but it was no use. He had disappeared into a sea of cars. In a moment, he had vanished as though into thin air. After five minutes, Miriam had called supermarket security. Then she had called Sergio at the pharmacy to tell him what had happened. After fifteen minutes, she called the police. The loudspeakers had notified everyone of the disappearance of a child. Alan's name rang out high and low. His face appeared on the maxi-screens and on each individual cart's display. A lot of people joined in the hunt. Other customers had opened the doors of their cars and looked in their luggage compartments, just in case he'd slipped in without their noticing.

Half an hour later, there was still no sign of Alan.

It wasn't until that afternoon, four hours later, that Alan was returned to Com-pro in a police car. The officers had gotten a phone call from a man who, annoyed by a loud car stereo, had come downstairs from his apartment on Via di Grotta Perfetta and found a boy unconscious in the back seat of a Mercedes. The child hadn't been able to open the door or release the power locking system but, before passing out from the heat, he had turned on the stereo and raised the volume as high as it would go.

Two policemen had pulled him out of the car in the nick of time. When Alan had revived, he had been frightened and began to cry. He knew his first name but not his last name, nor did he know his home address. Once they tracked down the car's owner, the officers found out that he had driven it to the supermarket, where they surmised the child must have climbed in, either by mistake or as a game. They had taken him back to Com-pro.

A car had imprisoned Alan and almost killed him, but music had saved him.

Miriam knew that her son had taken the job at Globalzon because corporate policy did not require employees to handle or interact with any sort of machinery apart from maneuvering a freight hoist – a piece of equipment that presented no threat to the dignity of human labor but which, by a cruel twist of fate, had been the very machine into whose arms he had fallen.

Only a few minutes have passed when Miriam's smartphone rings. "Hi, Ivan. You got my message?"

Miriam is reading through a discussion forum about vertebroplasty and kyphoplasty, techniques for stabilizing fractured vertebrae by injecting acrylic cement into them. She knows, from the reading she has done, that no way exists to regenerate a damaged spinal cord.

On the other end of the line, Ivan's words are brief and to the point.

"All right," she replies, "I know a place in Trastevere. A friend's family owns it.

"I'd like to go there."

★   ★   ★

Outside Il Romoletto, beneath the green neon sign spelling out the restaurant's name, a distinguished-looking man waits, smoking a cigarette. He wears a pair of round-rimmed glasses with dark lenses. He is bald and heavy of build, with sloping shoulders.

Ivan Mihailovich Shumalin is seventy years old. He's Russian by nationality, but he has lived in Rome for thirty years. Although he has a degree in medicine, he first came to Italy on a nurse's visa. He set a clear path for himself, working in the fields of Nutrition and Physiotherapeutic Neurology, and in time had the satisfaction of heading up eight-person research groups. Then, however, he had to suffer the bitter disappointment of seeing his career smashed in a head-on collision with the interests of the nutraceutical firms.

As soon as he sees Miriam approaching, Ivan puts out his cigarette and embraces her. She is wearing a subdued expression the likes of which he has never seen on her face before. He takes her by the hand and together they enter the restaurant.

Ivan is the talkative type, but Miriam's emotional exhaustion is such that their evening nonetheless risks becoming a silent one. There has been joy and intimacy between them in the past. It has never blossomed into a stable relationship but − for a while at least, a few years after her separation from Sergio − Miriam had a weakness for Ivan's intelligence and attentiveness. He had recently separated as well, from a Moldavian woman. Without ever actually talking about it, they had both toyed with the idea of finding a remedy for their sadness together. Then time and age had counseled them otherwise.

"I've been thinking about what you told me," Ivan begins, "and it reminds me of a time in my childhood, before penicillin. My mother was a nurse in a hospital in St. Petersburg. During the war she dealt with laborers, sailors, soldiers, drunks, refugees, and all of the illnesses they brought through the door. It was a perfect place to learn the basics of medicine. Every hallway was full. That is how many patients there were. Many had pneumonia. After five days, the bacteria would begin to multiply and spill from the lungs into the bloodstream. Within three days, their fevers went up to a hundred and four, a hundred and five degrees. There was nothing to be done. If a patient's skin remained hot and dry, it meant he would die. If, on the other hand, he sweated, it meant he would persevere, and while there were some sulfa drugs that had an effect on the mildest cases, the outcome of the fight against the lung infection depended solely on the patient's ability to endure it. My mother would come home exhausted, all of her strength spent. She told me over and over that a person's good health was one alone, but diseases were many."

The waitress who shows them to their table is a young, slim brunette with an absentminded air, who hurries to hand them their menus. Although it's Friday night, the restaurant – a traditional *trattoria*, with the requisite vintage photos on the walls and a genuine traditional horse-drawn cart from the eighteen-hundreds in the corner – is nearly empty. The only other customers are a couple of elderly Germans, who are listening to a Tourism Intelligence System relate the history of Ancient Rome over mugs of beer and *margherita* pizzas.

"When I came down with pneumonia, she knew, what with the fifty-percent mortality rate, there was not much she could do besides pray. It was while she was in church that a fellow nurse, one who had lost her husband in the war, told her that in Europe they had a magical powder that could cure the infection right away, in just a few hours. My mother was able to get a sample of penicillin sent to her through the hospital. They reproduced it in the laboratory, and so I was saved. Back then they called it a 'miracle of chemistry'."

"Could that happen for Alan? I know I'm asking a lot, Ivan. Believe me, I wouldn't have if the situation wasn't desperate."

Ivan is silent, but not for lack of an answer. Miriam's request could have serious consequences, not to mention side effects that are impossible to predict. He adjusts the red-and-white-checkered napkin in his lap. "Shall we order something?" As a rule, he is skeptical about the quality of food served in restaurants, but Italian dishes – their smells, flavors, and the way they are still served and presented – lure him in every time.

"I'm just here to keep you company," Miriam replies. "I'm not hungry. I'll gladly have a glass of wine, though."

"You invited me out so you could watch me eat?"

"No, so that I could listen to what you have to say."

After having worked for years in the field of regeneration, studying the abilities of certain animals – the salamander in particular – to regrow damaged or ruined parts of their bodies, Ivan had gone to work for the WFP. There, he had been assigned to a project called Ending Hunger, whose focus was combating malnutrition. With his nonconformist mentality, he had begun by reflecting on a few simple questions. Why did most of what humans eat have to be secreted back out through urine, feces, mucus, sebum, sweat, gas and blood? Why do some people, who eat healthily, often fall ill, while others, who eat nothing but junk, never get sick? Why do neither animals nor human beings feel hungry when they are ill? Why does the body react by proposing abstinence as the first action against any type of illness? The solution Ivan Shumalin had proposed was simple: stop eating.

"Alan says hello."

"How are things between the two of you?"

"It's as if I'm not his mother. I'm his nurse. Ever since they released him from the hospital, I've been living at his place, and I don't know when I'll be able to leave."

Ivan had developed the opinion that humankind should stop being obsessed with food and eliminate – or, at least, drastically reduce – that primordial need. It was a biological imperative to which nature had assigned a role of great importance, but at a high cost in terms of energy produced per gram of material consumed. It was also dangerous for the body because, in his belief, the majority of illnesses from which people suffered were dietary in origin.

When the waitress comes back to take their order, Ivan chooses chicken with peppers and a jug of the house white for them to share.

"Are you aware of the risks that we will *both* be taking?" he asks.

"Yes, and I want you to stay out of it. I'd rather run those risks alone."

"If that's what you want…. I've kept the plans for the Ending Hunger project, but I don't know if anyone will be able to help you – or if they'll do it for free. The community is open and sometimes people collaborate. Anyone could take that design and go in any number of directions with it, but what you're asking for is not easy to create."

"I'm not asking for a solution," Miriam says, opening her purse and pulling out a piece of paper. "All I want is hope."

Ivan recognizes that paper immediately. He has had to keep every such one over the years. That, however, is the first of them, printed on the old blue paper. When the jug of wine arrives, Ivan's eyes are bright with tears. Before him is the copy of his first residence permit, the one Miriam was able to get for him back in those frantic days after the FateBeneFratelli Hospital had refused to renew his contract. She had hired him, temporarily, as a caregiver for her son. That had been nearly twenty-five years ago. "There was no need for you to remind me of what I owe you."

"I didn't bring it so that I could collect on a debt. This piece of paper is for me, to give me the strength to move forward."

Ivan pours the wine for them both. "I've already taken steps. I've sounded out some of my contacts."

He pauses to allow the waitress to serve him his chicken with peppers. The meat looks tempting, and Ivan insists on tasting it immediately. He asks the girl to wait while he cuts off a piece of thigh meat and raises it to his lips. As soon as he puts it into his mouth, his expression changes. The taste is rancid, so revolting that he spits it out into his napkin. "This is absolute slop. Only tourists could eat it."

The Germans turn around, grumbling around mouthfuls of food.

It is with a grim tone that Ivan asks the waitress to try a bite of what she's brought him. She complies and immediately puts on an expression of disgust, perfect for placating an irritated customer.

"Bring me some roast potatoes, please."

The waitress, Carla – according to the tag pinned to her apron – retreats, plate in hand.

Who knows how many times she has to play out that little scene, Ivan wonders.

Less than a minute later, the restaurant's owner appears, a worn-looking woman wearing a faded shirt and a hairnet. She apologizes for the chicken, adding that the cook is trying to figure out whether the problem was caused by some sort of spice or perhaps the oil.

"My dear lady, that chicken is simply not fresh. I am, unfortunately, quite familiar with the odor of things which have spoiled."

"No, no, the poultry was delivered this very morning." Ivan looks at her askance. Her statement proves nothing.

Miriam signals Ivan to let it go. "That's all right, ma'am," she interjects, "the potatoes will do just fine. Won't they, Ivan?"

"Yes, all right. It appears this is not a good day for chickens." He flashes a sarcastic smile and takes a sip of wine while the woman retreats in embarrassment, eyes on the floor. "Oh, excuse me. I'd also like a portion of stuffed Ascolana olives while I'm waiting for those potatoes."

"I don't think you should be so hard on her, Ivan. All the more so because she must be Silvia's mother. You know, the girl who took me back and forth from San Camillo Hospital for weeks?"

"I'm sorry. I didn't intend to be rude. In the end, she might not even know in what state her goods are delivered. But believe me, Miriam, that chicken should never have been put on a plate. It had been dead for quite a long time. It should have been put in the ground."

"Her daughter would agree with you. Thanks to all those rickshaw rides, we've become friends."

"Is she the reason you wanted to come here?"

"Well, she told me that the ingredients were genuine, organic even. Knowing you, I thought it would be the right place."

"Sadly, what people have been saying about restaurants is true, and it doesn't surprise me. I hope they don't make the same mistake as those other *trattorie* that think they're five-star restaurants as soon as business starts going well."

"I think they've made mistakes like that before. Silvia has told me some unpleasant stories."

"At least they don't have musicians, painters, fortune tellers or some *maître d'hôtel* in a suit and tie."

Miriam lowers her voice and leans towards him. The problems affecting the restaurant business are not her priority at the moment. "So, when do you plan to do it?"

"As I've told you, I've already set things in motion. I have old friends in Moscow and they have some contacts…. Let's just say that, while they may not be the cleanest sources, they rarely make mistakes. I can send you the design proposal for the *nanites* as early as tonight. After that, you'll have to take care of the next steps yourself – that is, if you really don't want my help. You know I would be glad to give it."

"No, if I can handle the thing myself, then I'd rather do it alone. Is it truly that dangerous, Ivan?"

Carla comes back out of the kitchen. With shaking hands, she sets a plate containing both the potatoes and the Ascolana-style olives in front of Ivan, then stands there, waiting for him to approve. Ivan sorts out the olives and potatoes, which have gotten mixed on their way from the kitchen. He cuts a potato in half. This time, when he puts it in his mouth, he recognizes the proper texture.

"Excellent, they're cooked to perfection." Carla, gratified, hurries off to tell the owner.

Ivan chews slowly. His jaw moves in a mechanical rhythm. It's as though the flavors the kitchen has told him the food should possess have been divorced from any underlying pleasure, as if his palate were cut out of the equation, almost indifferent to the taste. It's not what he'd been expecting.

"You want to know if there will be risks, Miriam?"

"Yes, firstly for Alan and, secondly, for me." Miriam downs the rest of her wine.

"We're talking about the Silk Road, the Internet's hidden highway. Anything can, and does, happen there. Drug trafficking, arms trafficking, child trafficking…. But it's also the only place where you can find true experimentation, the kind that no corporation would finance, at least not in the light of day. It's there that the knowledge shared by anonymous users gives rise to the inventions of tomorrow. I'm sure that there will be many people anxious to get their hands on the design for those nanites, people who have the means to develop them the way you want them to. I cannot. I would need a Software Agent capable of tapping into the computing power of thousands of servers, the kind used for distributed computing projects. Even if I found a way to get my hands on all that power, I would need a long time, too much time – if we don't want to commit some kind of crime, that is. And your son doesn't have five or ten years to throw away."

"Ivan, you didn't answer my question."

His reticence is not necessarily a bad sign. Rather, it shows her that Ivan doesn't want to involve her in any shady dealings.

"As I've told you, you'll have to offer the design for the nanites in exchange for the development of a prototype that will restore the functions of Alan's spinal cord, which is not like asking someone to make you some Viagra."

Ivan cuts an olive neatly in half. "Here. Let's say that the oil from this olive represents spinal cord tissue, which Alan needs." He points to the inside cavity with the tip of his knife, then scrapes it along the inner wall. "But he's missing the stone, and without it the fruit cannot grow."

He pops the olive into his mouth. "The nanites are the missing stone."

CHAPTER THREE

Rainbow Magic Land

A hand grabs Miriam's wrist, gripping her tightly.

"My legs! My legs are gone! They're not there!" Alan wakes with a start, as he often has since he came home from the hospital. It's been three months since the accident and roughly two since Miriam's meeting with Ivan.

"No, they're there, even if you can't feel them." She lifts the sheet. "See? Everything's still there. It was just a dream." Miriam looks at her son, lying in bed with a growth of unshaven beard, his eyes red and a dark expression on his face.

"I'll never work again...."

He's been wearing the same Rage Against the Machine shirt for the last six days. She hasn't been able to get him to take it off.

"Don't say that. I got an email today."

Things have been set in motion, things Alan knows nothing about and that Miriam fears.

"No, it's true. I'll never work again."

Miriam makes an effort not to look at the sheets and the motionless contours of Alan's legs beneath them – the legs he once used to run, jump and, when he was a boy, to play basketball.

"Now that I'm an invalid, with legally protected status, it would be easy to find a good job, but I don't have any intention of working for anyone, ever again, not for any reason in the world."

"I just told you I got an email. It's something important."

Alan isn't listening to what Miriam is saying. She can't really blame him – not only because of what happened at Globalzon, but also due to what he went through before that. He and his generation have had to accept jobs answering telephones, delivering mail and packages, filling up cars at the gas station, working as private tutors, serving hamburgers and French fries.

"Help me sit up."

Miriam presses the button that raises the special bed she has had installed in the house.

"Not like that! With your hands!" Alan can barely contain his rage. He leans over towards the trunks of his legs and pokes at different parts of them. "You tell me, Mom. What's the point? If I was a burden before, now I'm a worse one and I will be forever." Miriam doesn't even listen to her son's ravings. Instead she hands him a glass of water and wonders if he'll appreciate what she's about to show him. On the one hand, she understands his despair and frustration, but on the other, she hopes that they have not already turned into acceptance and resignation, because she intends to help him snap out of it and fight back.

"Listen, Alan. There's a video I'd like to show you. But you have to promise me you won't interrupt or make any comments until you've seen the whole thing. All right?"

He sneers, but he agrees.

Miriam pulls her smartphone from her purse and opens up the email. She clicks on the video it contains and plays it for her son. It summarizes the different applications of nanites in the medical field: blood oxygenation, wound repair, bone reinforcement.

Alan's eyes widen, his curiosity aroused. Although the visual representation of the nanites is like that of microscopic motors capable of assembling matter atom by atom, in the video they do not appear to behave like machines. On the contrary, their actions imitate the natural processes of molecular production, reminding Alan of high school chemistry lessons.

When the clip ends, Miriam takes back her smartphone and puts it back in her purse. "These ideas may never see the light of day. They are projects that require backers, sponsors and goodness knows how many ministerial approvals, not to mention authorizations from multinationals."

"Then why are you telling me this?"

"Because I wanted to see your reaction. I know about your aversion to machines."

"Things change. Besides, those nanites look more like molecules than machines."

Miriam is pleased to see that her son hasn't lost his clarity, even in the midst of his tragedy. "In that case, I'd like to talk to you about certain... clandestine projects. Alternate ways for people to acquire things that they want."

"Be clearer. What are you up to?"

She lowers her eyes, ashamed, even though her actions have stemmed only from love, a mother's love. "I've done something wrong, committed a crime. But I may have found a way to make you walk again."

"How? What have you done?"

"I can't tell you any more right now. Just...have faith and don't despair."

Alan is mystified. His mother has never been a courageous woman. When she would call him on his cell to find out who he was with and what he was doing, it was more to reassure herself than for his sake. In her whole life she's never taken any big risks. Besides, she works at the WFP, with one of the last permanent contract positions ever drawn up in Italy.

The only real shock in Miriam's life has been her separation from Sergio Cormani, and that was consensual, free of the drawn-out friction and emotional backlash such events usually entail. Sergio had been a pharmacist, like his father and grandfather before him. Naturally, he had wished for Alan to follow the same path. The Cormani family had always been a point of reference in the neighborhood between the Via Appia and the Via Tuscolana. Everyone knew who to go to for advice about pills, salves and prescriptions, without always having to consult their family doctors. However, after the umpteenth argument between father and son, arising from the difficulty of finding and – more importantly – being able to hang on to a scrap of work, Miriam had chosen to take Alan's side. Sergio had not understood how someone could simply throw away a tradition that had lasted nearly a hundred years.

Alan was wearing headphones on the day Sergio left. Whenever the pair were forced to meet after that, on the occasional family holiday, their conversations never went on for any longer than the length of a few songs.

It was after the divorce that the true reason for Sergio's sudden departure became apparent. He had been seeing a customer of his, a woman who lived in Lanuvio, a little seaside town outside of Rome. Once he had moved there to be with her, all father–son relations had ceased.

"So, I'll have to come visit you in jail?"

Miriam puts on her jacket and goes over to embrace her son. "I hope that in just a few days I'll be able to give you some good news. For now, though, I'm off to run some errands."

"Don't make me worry about you. It's enough that one of us needs taking care of."

Miriam plugs Alan's smartphone in to charge. She feels a sort of strange strength inside of herself, the kind that comes from undertaking something that, though terrible, is also extraordinary.

That afternoon Miriam receives instructions on where and how to meet up with her Silk Road contact. As far as she knows, it could be anyone – one of those people from Anonymous, a political activist from Occupy the World or even some delinquent from Pirate Bay, all 'hacktivists', as they like to call themselves, all operating in the shadowy places on the Internet, in the borderlands between legality and criminality.

Strange fact number one: the Bartolini courier doesn't ask her to sign any notice of delivery, and the package he leaves has no markings.

Strange fact number two: the package contains an old-fashioned cell phone, an old flip model.

Strange fact number three: a Post-it note that is stuck to the phone. It reads:

TURN ON PHONE
READ TEXT MESSAGE IN 'SAVED' FOLDER
DELETE AFTER READING

Once Miriam has followed the instructions – the message sender signs himself 'St. Anonymous' – Miriam checks the time and then her reflection in the mirror. She pulls her gray hair back into a braid and puts on a lightweight orange shirt and pistachio-colored cotton pants. Onto her feet go a pair of cork-soled sandals.

"Alan, I've got to go out to run some errands. I may be a while. If you need something, call Carolina. I've left her a copy of the keys."

"Where are you going dressed like that?"

"To the park."

"What are you going there for?"

"For now, I'm going there to walk. For you."

Miriam leaves the house and heads for the San Giovanni metro station.

The last free shuttle for Valmontone leaves from Via Marsala, next to the Termini train station, at five p.m. It will take about forty minutes to reach its destination.

Miriam uses her smartphone to buy an evening ticket, valid until eleven p.m., even though she hopes to be home before then. While boarding, she looks carefully at the other passengers in line at the doors. She walks slowly down the shuttle aisle between rows of seats, looking for a free spot but, more importantly, for some useful clue.

The bus is full of children and the din is deafening. Their shouts and scuffles drown out both the traffic noise and the sound of cartoons being shown on the monitors.

It has been a long time since Miriam last went to an amusement park.

When Alan was little, the old Luna Park in the EUR neighborhood had still been open. That had been the only place where Sergio was willing to spend a few hours with his son before going back behind his pharmacy counter. Later on, the public amusement parks and playgrounds had disappeared, until all that remained were a sprinkling of slides, sandboxes and some swings around the city's various neighborhoods. Large-scale private theme parks, however – immense cathedrals of lights, scents and excitement – had grown in number and in size.

Miriam finds a place to sit and opens the brochure. This year, Rainbow Magic Land is celebrating its twentieth anniversary. Since its grand opening, it has expanded to cover an area spanning several hectares. Today, between service and maintenance technicians, entertainers and seasonal employees, it provides jobs for 3,274 people. The economies of numerous towns on Rome's outskirts, from Colleferro to Palestrina, from Artena to Valmontone, depend on activities linked to the park's operation.

"Hands up!"

A blond boy, a little scamp no more than six years old, is pointing a water rifle at Miriam's forehead. She obeys, and the boy leans forward from the rear seat. His mother grabs him by the shirt – the same way she's been holding two other children sitting next to her – but he resists.

"Why are you by yourself?" he asks.

Miriam becomes aware that she's the only one on the shuttle who doesn't have any children with her. The realization that she stands out makes her uneasy. "I'm not going by myself. My grandchildren are waiting for me at the park entrance."

The imp turns away with a pout, then squirts his younger siblings with a liberating jet of water.

In Rome there are roads that start in one century and end in another. Some span millennia. Via Appia, the old Appian Way, is one of the latter.

Miriam watches as the buildings facing Piazza Re di Roma, their solar-energy balconies and façades covered with sound-absorbing resins, give way to buildings dating from the reign of King Umberto I, decorated with friezes and stucco relief-work and enveloped in ivy. These are succeeded by cheap decaying apartment blocks looking out over the train tracks, which in turn are replaced by houses in a state of semi-collapse, propped up against the piers of the Roman aqueduct. Next comes the Grande Raccordo Anulare, the Great Ring Road around Rome, followed by an encircling belt of hills and, finally, the asphalt strip of the highway.

Getting off the bus half an hour later, Miriam recalls the message she read on the cell phone:

RAINBOW MAGIC LAND

SIX EYES: TWO LARGE AND FOUR SMALL LOOKING EASTWARD

The riddle is an imaginative way of putting her on the trail of her Silk Road contact.

Under the rainbow that serves as the entrance to the amusement park winds a line of people easily a hundred yards long. Dozens of buses from various cities are unloading families in an ongoing cycle. Waiting for them in an open lot crowded with teepees, livestock trailers and stagecoaches are two groups, one of 'cowboys' and another of 'Indians', played by Slavs and Indians (from India), respectively. The 'Native Americans' are tall and grim, their faces painted red, white and black. Their hands grasp the hafts of rubber tomahawks and they are naked but for short suede loincloths and leather sandals. Meanwhile, the cowboys flaunt absurd mirrored sunglasses and leather vests beneath their wide-brimmed hats. The tips of slant-heeled riding boots peek out from beneath the hems of their studded bell-bottoms.

The 'cowboys' ride long-haired ponies in a circle formation around the Indians. Their heads bob up and down and every now and again they yell out a resounding *yahoo*. The 'Indians' are impassive, paying no attention to the cowboys. At most, they occasionally exchange slaps on the shoulders, as though trying to keep each other awake. Behind them a mariachi band plays a selection of cheerful Mexican 'folk' tunes: 'La Bamba', 'El Carretero', 'La Cucaracha', 'Cielito Lindo'.

Miriam gets in line, looking at as many people as possible. There are lone parents with lone children, little groups of grandparents tasked with managing the grandchildren, couples worn out by their children's demands and nannies armed with patience and sympathy.

A little apart from the rest, as though waiting for someone, is a woman with Asian features. She is standing still in front of the beverage kiosk holding the hands of two Filipino children. Miriam focuses on them. The woman does not appear to be the children's mother. Her bearing is stern, too rigid and in no way maternal. She is playing her role poorly, wearing an outfit more suited to the office than to taking one's children to the amusement park.

Six eyes, two large and four small, looking eastward.

Miriam approaches them. Doing so will get her in among the families. That way, in any case, she won't attract anyone's attention. "Hello…. Have you been waiting long?"

After a long and ceremonious smile, the woman takes a smartphone from her jacket and types something out with impressive speed. Then she shows Miriam the message.

THIS IS HOW WE WILL TALK. YOU READ MY TEXT THEN I ERASE IT. YOU DO THE SAME WITH ME

Miriam nods and follows the group as they head to the ticket booth.

"Well? Where do you want to start?" The woman is talking to the children, who answer in unison without a moment's hesitation.

"Huntik 5D!"

To play Huntik 5D – as shown on the ride's display screen – you have to climb into a special car, put on an enhanced-reality visor and arm yourself with a laser pistol for shooting at the monsters that appear along your route.

The children, in a state of high excitement, take the front seats, while the two women sit in the back.

Once in motion, the car accelerates gently around the first curve and then moves forward into darkness. Shouts of fear and happiness mingle as the lasers hit betentacled aliens, spinning spacecraft and other unidentified targets.

The woman begins to type messages like a pianist on a keyboard.

GIFT GIVEN IS VERY VALUABLE.

REQUESTED PROTOTYPE FOR YOUR SON DEVELOPED IN EXCHANGE.

Miriam responds immediately.

WHO ARE YOU? WHAT DO YOU DO?

The woman's answer requires more time to type out.

WE ARE ENGINEERS DISSIDENTS EXPATRIATES FROM CHINA TO EUROPE. LIVE WITH HUSBAND IN PISA. MAKE SHOES, 3D DESIGN ITALIAN SHOES. YOU KNOW MODS?

Dodging a blow from a bloody zombie covered in cuts and scars, Miriam ducks down to type her answer.

OK BUT WHAT DOES THAT HAVE TO DO WITH NANITES?

The woman gives the two snipers in the front seat some words of advice then resumes typing.

FOR YEARS WE DEVELOP NANOTECHNOLOGIES: WATER FILTERING, ALTERNATIVE ENERGY SOURCES, HEALTH AND LONGEVITY.

Miriam's second question is of a more delicate nature.

WHAT DO YOU PLAN TO DO WITH THE DESIGN?

Once more the wait lasts several seconds.

HUSBAND SUFFERED HUNGER AS A BOY, THEN BECAME RED GUARD. AFTER 1969 WAS SENT TO COUNTRYSIDE. SUFFERED HUNGER AGAIN. WE HAVE ONE PURPOSE: CREATE NANITES TO MAKE VITAMINS, PROTEINS AND FATS FROM RAW MATERIALS, FROM SIMPLE MOLECULES THAT EXIST IN NATURE, WATER, AIR.

Miriam is reminded of Ivan's project, Ending Hunger. A lot of people have all been rowing in the same direction. Then she imagines an old man, an infallible *sensei* of molecular composition, to whom experts the world over would send their 3D plans, all waiting for the master to choose a project then turn his infinite patience and atomic precision to its completion. It was a bit like building objects using only grains of rice. The molecular industry had restored the craftsman's trade.

She remembers reading an interesting article a few years back. Die-hard Lego fans had been sending the company endless suggestions on how to improve and customize the design of their Mindstorm Robotic products. Sergio had tried to give the young Alan some of them once, in the vain hope he would develop a passion for the concept of molecule/brick. The article had gone on to recount how Lego had initially intended to sue its own fans for unfair competition but then thought better of it. It had made an announcement on its website, saying it was introducing the 'right to modify' as part of the Mindstorm software license, thus granting its devotees permission to use their own imagination on its products. When Lego Factory had opened – a 3D virtual workshop where users could create, share and sell their own customized bricks – customers were able to unleash their creativity with no fear of being reported to the authorities.

Every now and again a threatening presence appears out of the darkness of the tunnel, a three-headed monster lying in wait, a fearsome alien in a gold jumpsuit or a snarling murderer, but all are put down in a heartbeat by the lasers of the young avengers.

Miriam's next question is already hovering at her fingertips.

HOW DO I GET THEM?

The woman puts a hand into her jacket pocket and pulls out a memory stick, which she places in Miriam's purse.

I PASS YOU FORMULA NOW. PRINT AT ANY PMC THEN GIVE TO YOUR SON TO EAT. MAY HAVE SIDE EFFECTS. WE TESTED PROTOTYPE ON TEST SUBJECTS FOR SHORT TIME. RESULTS ENCOURAGING. VERY!

Thinking about the risks Ivan spoke of, Miriam formulates her next question. She wants to understand the reason for so much secrecy.

HOW MANY DIFFERENT PEOPLE WILL BE LOOKING FOR ME?

The woman's lip curls up in a hint of a sneer, then falls, as though it is not in her nature to be cruel.

COPYRIGHT OWNER FOR SURE. PLUS, WE TOOK COMPUTING POWER FROM MILLIONS OF PCS. BUT THAT IS NOT A PROBLEM. RAN SOFTWARE AGENT ON SLEEPING COMPUTERS IN OFFICES OF 'EVIL' MULTINATIONALS.

Miriam is more worried than before. On her cell she types:

WHICH ARE THOSE?

The answer is immediate.

YOU KNOW. JUST GOOGLE FOR UPDATED LIST. IF THEY COMPLAIN ABOUT INTRUSION, WE HAVE SKELETONS TO PULL OUT OF THEIR DATABASE. NOT WORTH IT FOR THEM.

When they get to the end of the ride, Miriam says goodbye to the children and the woman and heads back towards the shuttle. Suddenly she turns around and goes back the way she came. She has one last question to ask.

WHAT RAW MATERIALS DO I NEED?

The answer is immediate.

SIMPLE CARBON CARTRIDGE. JUST LIKE WITH STEM CELLS, THE DIFFERENCE IS IN THE BLUEPRINT.

CHAPTER FOUR

PMC (Public Matter Compositor)

The line for the PMC on Viale Marco Polo stretches all the way out to the sidewalk. For years, Matter Compositors have been used for industrial prototyping, but ever since they first began to appear outside of factories, many people have taken to using them for the most disparate of reasons: young couples reprinting dishware destroyed during a domestic argument, elderly people who need new walkers, glasses or prostheses for daily use or girls looking to catch someone's eye using the latest fashion accessory – decorated nails, phosphorescent teeth or subcutaneous inserts intended to provoke an intense emotional reaction.

Users' imaginations know no limits, and the range of possibilities offered by design software, in combination with the 'social designs' that can be found everywhere – from forums like Wik-It-Self to sharing networks such as Just-Nano-Do-It and Have A Fab! – supply the perfect response to a growing desire for creativity for its own sake. On the occasions when this leads to a global success, it is often in the form of money raised from Online Auctions, where patents are the object of heated bidding wars.

Miriam, who is stuck in line between a mother whose child is crying over a broken toy and a repairman resigned to waiting to print out some electrical sockets and construction material, feels anxious. The memory stick is gripped tightly in her hand, as though it were the most precious object in the world. From a mother's standpoint, it is.

Next to her is Ivan, who insisted on accompanying her.

The PMC staff are moving the flow of people along as quickly as they can, but every once in a while some pimply-faced kid ties up the machine with a particularly tough composition. What's holding things up at the moment is a mountain bike in ultralight carbon fiber. The kid, confident that the end result will be worth a few insults and a couple of shoves, couldn't care less about the crowd behind him or the racket around him. Behind him, an old man is waving a set of dentures that need fixing.

"Everything will be fine, you'll see. If our friends have done their job, Alan will be able to walk again in a matter of weeks."

"Do you believe in miracles, Ivan?"

Miriam has a talent for shifting the conversation on to more serious topics than he had been expecting.

"Years ago, before I came to Italy, I used to go and pray. There's an old church in St. Petersburg called the Church of the Smolensk Icon of the Mother of God. Many people go there to make vows to St. Xenia. It is said that the she transported all the bricks for the building of her church to the site in a single night. Don't you think that's a curious coincidence?"

"And what did you ask her for?"

"I asked for a wife and a family."

"So, then, you don't believe?"

"Maybe it wasn't such an important thing, after all. From a certain point of view, you became my miracle worker when you helped me."

Miriam slips her arm through Ivan's. "Oh, come on, you're comparing me to a saint now?"

"No. I was trying to say that it's not always necessary to be a saint to do extraordinary things."

Once inside, Miriam and Ivan find themselves looking at the latest model of Matter Compositor. It has the same dimensions as a shipping container, so it can be transported by truck. Its users are not daunted by its size. The ratio between human- and industrial-scale feels properly balanced. Its walls are transparent, so most of the interior is visible. It is filled with vats of raw material, cooling systems and backup power-supply units.

Ivan, who is beginning to feel excited, pushes Miriam closer, jostling a muscle-bound tough who turns around, irritated, showing off the bar he intends to load up with new heavy weights. "Hey man, this isn't a mosh pit."

Ivan doesn't lose his composure. He apologizes, then ignores him, turning his attention back to Miriam. "You see, my dear, this gadget could provide supplies for any neighborhood in Rome for six months, without needing a single refill."

There's something magical about the machine, something alchemical – and these days, alchemy is synonymous with nanotechnology. The price per kilo for the raw material ranges from a maximum of four hundred and fifty euros to a minimum of seventy, depending on the design complexity and assembly time.

A heated debate has been raging online for a few months now about the possible consequences if a PMC were to be installed in every home. This has happened before, first with TVs, and later with PCs and cell phones. In the case of the PMC, some are concerned that factories would disappear altogether, some worry about the demise of the shipping industry, while others fear the collapse of the industrial economy. Then there are those who, on the other hand, hope for those things to come to pass, heralding the transition from a market economy to one of free trade or, better yet, some form of trade that is truly free from the brokerage of obscure entities such as banks.

When it is finally their turn, Miriam hesitates. "Insert the memory stick, like so."

She follows Ivan's instructions, and a message appears on the display:

COPYRIGHT VERIFICATION IN PROGRESS
SEVERAL MINUTES MAY BE REQUIRED TO COMPLETE THIS OPERATION PLEASE WAIT

Ivan places his hand on Miriam's shoulder. "Don't worry. I took care of it."

"But, I asked you not to—"

"I owe you that much, Miriam. I simply asked some old Russian friends for a favor."

"I know about your 'friends'. They're the ones who steal information from databases and sell it to the highest bidder."

Ivan feigns offense. "How can you say such a thing? They're top-notch computer security consultants."

"By day. By night they're hackers."

"Krypton, my childhood friend from the *Konsomol*, would not take it very well if he heard you make such insinuations. We were Young Pioneers together. We played in a rock band in the nineties. We used to hang out in the last of the Dom Kultur centers. Krypton was happy to help your son by validating the certificate for *my* design."

"Are you sure? If something bad happens, 'they' will come after you."

Ivan shrugs. The Ending Hunger project had been based on an idea he had worked on for years. In 2019, fifty-two teams had taken part in NanoBlock, an international competition whose purpose had been to assemble molecular sequences capable of producing enzymes and physical properties. Participants had received NanoBlock kits, each of which included a pack of interchangeable standardized nanosomes. All modifications made by the teams were saved and recorded in a special biological sequences archive, an online database which functioned according to the principle of 'take one, leave one', so as to encourage the exchange of information. Ivan's team had taken first prize, a hundred-thousand-dollar grant, no strings attached, to cover the cost of continuing their research for the next two years. The final stage would have revolutionized the food industry, eliminating the intermediate phases of production, distribution and storage, not to mention the disposal of waste goods and foodstuffs, but when the time had come to move on to that last step, the sponsors had disappeared, either cowed by threats or bought by the latest nutraceutical companies.

"The important thing is that nothing can lead them to you. Besides, what could 'they', as you call them, take from me? My car? My house? My job? I'm seventy years old, Miriam. Not many things in this life have a hold on me anymore."

A new message appears on the screen:

SELECT DESIRED QUANTITY AND COMPOSITION SPEED

From the dropdown menu below, Miriam selects:

QUANTITY: 3   SPEED: MAX

"Why three?"

"I'm not sure that Alan will be able to do it the first time. And I want the third copy to be for you, just in case of anything."

Miriam takes a deep breath and looks at Ivan. In her eyes is the fear felt by someone who is about to set off a series of reactions whose consequences are impossible to predict.

"It will happen sooner or later in any case, my dear. Besides, you're doing it for Alan. Someone will take that into account."

The moment Miriam touches the screen, the PMC begins to composite the design. Countless carbon atoms stream from the nozzles lining the axes X, Y and Z and are deposited one atop the next in a predetermined pattern. The screen displays a real-time 3D graphic simulation. Layer by layer, in the form of the finest of pollens, the design becomes solid.

Ivan is captivated by that vision. The nanites' existence would be bound to that of their host body. Indeed, in a symbiotic fashion, they would depend on their host for their very survival. From that perspective, they were similar to common parasites, although they had been designed to improve and strengthen the organism rather than exploit and weaken it.

"Are those nanites 'alive'? With the risks you told me about?"

"Alive is a strong word for it, but it's a fine approximation. With time, swarms of nanites could develop their own emergent behaviors, exactly like chromosomes do when they transmit genetic data: the information they contain is hereditary, but it can also be modified through interactions with other genes. As far as the risk of uncontrolled proliferation is concerned, I'm skeptical...."

"They'll just keep on cloning themselves?"

"I'm not afraid of cloning. I believe that our knowledge and education will continue to differentiate us from each other. Our culture will save us from homogeneity."

"But the opposite risk exists, too," Miriam responds. "Look what's happening today. Cultural cloning is far ahead of the biological sort. It works through ideas, lifestyles, consumer preferences."

"You're right, and yet no living being endowed with even a glimmer of intelligence would be so stupid as to self-destruct, destroying its own habitat."

"Then what of man?"

"My point exactly. Man is always the danger, because right now the nanites' intelligence is still the product of human programming, of the magic fingertips of our Asian friend. All we need to prevent the sort of disasters of mythic proportions prophesied in reams of doomsayer literature or scenarios of death and destruction where the world ends up being swallowed up by billions of tiny, voracious, tireless little machines is a set of basic rules. In my opinion, if mankind hasn't already been wiped off the face of the Earth by the Great Flood, the Black Death or world wars, or by the atomic bomb, AIDS, the hole in the ozone layer or robots – well, then it won't be destroyed by nanites, either."

"What's happening now?"

The PMC has almost finished its composition of the first nanite.

"The printer is distributing the final molecules of your 'seed' nanite at a rate of one hundred and six atoms per second. The first nanite can reproduce itself without any outside help. It will guide the composition of other nanites, which will begin to differentiate themselves from each other in order to produce the desired effect."

"What effect, Ivan? Alan's vertebrae are broken."

"Healthy bodies are complex systems. You know, for instance – or, at least, your body knows – when you're eating too much or too little. In the same way, you know what is right or wrong for your body on a basic organic and biological level. However, you have no awareness of this knowledge, nor can you control it. You don't know how long it will take for a headache to pass or for a wound to heal. You don't even know – at first glance – whether what you are eating is good or bad for you. You figure it out by trial and error, through experience, through education.

"With the nanites, it's different. Their speed and efficiency are superior to that of organic cells." Ivan brings one hand down atop the other in a chopping motion, then slowly brings the fingertips of both together until they touch. "Peripheral nerve fibers do regrow. The Schwann cells lengthen by a few centimeters, allowing nerve signals to be restored. Otherwise, every time you got a cut on your finger, you'd lose all feeling there. However – and we have yet to figure out why this is true – neurons and central nervous system fibers aren't able to grow back. The spinal cord is not capable of lengthening and reconnecting, not even by one centimeter, but it does happen – and in a marvelous fashion, too – in salamanders and some other animals. Inside their bodies, the ends of the sundered segments seek each other out, forming spiral segments called 'neuromata'. The essential thing, however – and this is true in Alan's case, as well – is that the cells underlying the break must not be dead. The reflex arc must be intact, so that the nanites we are compositing will be able to do for the spinal cord what Schwann cells do for the peripheral nervous system. Think about it. It's in their interest for the host organism to be healthy. Parasites want to live too, after all."

Miriam doesn't appear as cheerful as Ivan. "So they're really parasites?"

"They're not dust mites, nor do they carry diseases. They are something that has been improved by observing nature. They are the technological evolution of stem cells and chromosomes, which is why some people go so far as to call them *nanosomes*."

"Nanosomes…which could alter the DNA."

"No more than genes do already. DNA is not an immutable thing. On the contrary, it is constantly changing over time. It's the idea of another human tampering with such an intimate part of ourselves that upsets the small-minded – the same people who don't want to recognize the fact that even our daily existence alters the structure of our DNA. Training our bodies and educating our minds – these are activities connected to the structure of our genetic code, but no one has ever dreamed of saying that they corrupt human nature. If anything, they say the opposite."

A message appears on the display.

PRODUCT COMPOSITION COMPLETE SELECT YOUR PREFERRED DELIVERY MODE: LIQUID, SOLID, GAS

"How does liquid sound? Maybe I could mix it with something else, make it easier for him to swallow."

"Good idea."

Miriam touches the screen and, after a few seconds, a transparent liquid begins to drip into a one-and-a-half-ounce phial.

Ivan's mind can't stop processing what he's observing around him. To see so many people in line here – it's a symptom of change. "All of this, our myths, our culture, our rules, our limits, the rhythms of our lives...we create every aspect of our existence, so why not take the next step?" It's not the finest of metaphors, but Ivan fails to notice.

"That's what I'm planning to do, if Alan cooperates."

"Plus, we're always torn between two ideas. The first is that the mind is like a blank page where our families, our education and our personal experiences imprint their golden rules. The second is that there is a sort of individual destiny written into each of our genetic codes. So, you see, nanites do no more than impress culture, health and education into our DNA, but instead of doing it through genetics, they do it memetically. A meme is like something passed on by word of mouth, a story that is transmitted freely from person to person because it works. It's an intangible replicator of information and behavior."

"I hope I can remember how to put the same spin on it when I'm talking to Alan." As soon as she removes the first phial, the second begins to fill.

"It's not a spin, Miriam. Besides, your son's forty years old. He should be able to take a bitter pill without a spoonful of sugar."

Miriam chooses to ignore Ivan's last admonition, which seems, in part, a comment on the way she has raised Alan. She wraps the phials in a cloth and places it in a Tupperware container she's brought from home. The bill emerges from a slit in the PMC.

MATERIALS COST = 5 EUROS DATA PROCESSING = 25 EUROS
TOTAL = 30 EUROS
SELECT PREFERRED PAYMENT METHOD

Miriam chooses to pay in cash. She slips a €20 then a €10 bill into the slot and waits for the payment to complete. Finally, she and Ivan leave the PMC center. Outside, dozens of people are still waiting to composite their dreams.

"I can't believe it. Alan's salvation at such a low price?" Ivan doesn't answer. The pair head back towards the car.

CHAPTER FIVE

Bucatini

Today's meal is Alan's favorite.

In the time since she returned home, Miriam has been able to sidestep her son's persistent questions. When will he get to see these mysterious nanites? How will he ingest them? How long will it be before he can walk again?

Miriam has stalled, giving nothing away and maintaining a cautious silence when it comes to her 'missions', first to acquire and then to print the nanites. As Miriam is draining the pasta, Alan bumps into a doorjamb and begins to shout, spewing out a stream of curses.

"Stupid fucking machine! I hate you!"

He still hasn't learned how to maneuver the wheelchair, or rather, his fingers refuse to manipulate the joystick. The electric mini-scooter is a wheeled 'mobility device' that looks like a cross between an executive office chair and a golf cart.

"Come on in, Alan. I've made something you like and I've put in a surprise ingredient."

Alan appears in the kitchen doorway, glowering. He brakes suddenly and the chair slides across the tiles. He maneuvers himself over to the table, driving his chair so far underneath it that his chest bumps against its wooden edge.

"What did you do to your hair, Mom?"

"Nothing. I let it down is all."

"Maybe I don't see too well anymore, but it looks so…shiny."

"Maybe you spend too much time stuck in the house."

He scoffs and strains to peer as best he can over the tabletop.

"Where would I go?" He leans forward over the plate, looks at its contents and scowls. "So, these are the famous nanites? No, wait, don't tell me. They're not *bucatini* at all! They're the incredible, all-new *bucananites!*"

Miriam lifts the steaming plate and offers it to him. It contains four ounces of *bucatini all'amatriciana*. Somewhere in the enhanced sauce, invisible to the naked eye, swim the nanites.

"You're in a good mood. That's for the best, because you may be able to walk again, but there's a price to pay, and a chance they may not work at all."

Alan is wary and a lack of assurances is making him skeptical. When he touches his legs, he has the bizarre feeling that he is stroking the bark of a tree. His limbs are alive inside, even if they are incapable of movement.

"So, what do I do? Do I just eat them?"

"Yes, you *have to* eat them, Alan."

Alan reaches for his silverware, his brow furrowed. The concept is simple enough, but the implications are profound and beyond imagining. Still, he is out of choices. It's either prosthetics or an exoskeleton – both of which he refuses – or the nanites. The third option is not to choose at all, and remain planted like a vegetable in his wheelchair.

Miriam extends the plate another inch towards him. "Come on, Alan. Your legs are waiting."

The way she says it, it's like his legs are sitting there on the plate in front of him. Alan lifts his fork. It feels so heavy, a weight that, once he lifts it, could be capable of putting his body and his entire life back into motion.

"Do you really believe it? Or are you just trying to make me feel better?"

The first attempts to bring his legs back to life failed. The doctors at the Santa Lucia Institute placed weights on his thighs, lowering them from above with little pulleys. Four, eight, twelve, sixteen pounds, but always nothing. No contraction, no reaction. He had wanted nothing more than to feel pain. Even a little would have been a sign that there was still just a tiny bit of life left in his legs.

Staring at the plate, Alan imagines that the nanites are a sort of fine ash, the sort that gets everywhere, into everything. Something dirty. Swarms of them, the consistency of dust, depositing themselves along the lengths of his bones and on his organs, surging through his bloodstream and claiming his stomach for their own, colonizing him from the inside – or else, a symbiotic salvation.

"What if these nanites kill me instead of healing me?"

Miriam takes a seat. The persuasion stage has arrived and she feels prepared. "That won't happen. They're like genes. They live as long as their host is alive."

"What if they don't work? What if they make me deformed, some sort of mutant monster?"

Miriam has an excellent answer to that question, but she doesn't want to waste it just yet, not when she doesn't know how long it will take to convince him to *dare* instead of *fear*. "The nanite prototype was tested by the person who gave it to me."

"Tested? On whom? For how long?"

"On mice, and for long enough. The people I got it from aren't amateurs, if that makes you feel any better."

"Mice? And that's enough for you to go ahead and feed them to me?"

Miriam stops answering. The things Alan is asking aren't actual questions. "Would you prefer me to leave? Would you rather do this alone?"

Her son is the sort of person who has to believe the decision is his to make. "How long before they take effect?"

"Two weeks. That you can be sure of."

Alan lifts his fork and slowly begins rolling himself a mouthful of *bucatini*. Miriam gets up and leaves her son to continue on his own. There are things he doesn't know, things he'll come to understand over the next few days.

He pauses before taking a bite, the fork poised in the air. He turns to Miriam and offers a hint of a smile. "What about side effects?"

"Eat, Alan. The pasta's getting cold."

★   ★   ★

"Mom! Mom, help!"

The last time Miriam heard Alan call out to her for help like that was at the hospital, after he'd hit his head against the corner of a wall in elementary school. He had been seven at the time. They had rushed him to the emergency room, his forehead covered in blood. She had held his hand while they sewed it up – more screaming and sixteen stitches, still visible just above his hairline.

Miriam hurries into the room next to her own. "What is it, Alan? What's happened?"

"I feel awful. I don't have any strength left at all. Please, give me some water."

This time, instead of being focused in one area, the pain is spread throughout his body, deep and ubiquitous. Miriam has moved into his apartment, the better to assist him, and she is thinking of renting out her own in order to meet her growing expenses. Physical therapy and painkillers are draining away all her savings. Miriam has been on leave for two months now, and Cecile isn't happy about it. She's put up with it thus far, but how long will that last?

Alan's eyes are fever-bright and his complexion is ashen. His water bottle is empty.

Ever since he ingested the nanites, Alan has been drinking more than three liters of water a day, but he hardly ever uses the bathroom. Most of the liquid he takes in he just sweats back out. The nanites are absorbing a large amount of his body's energy.

They are likely to be reproducing at an extremely high rate as they deploy and will do so until they reach the number that is optimal for her son's body.

Alan grabs suddenly at his own elbow. "There's something here, inside."

"Trust me, there's nothing."

He hasn't grown gaunt, but the muscles in his arms have become thinner and somehow deflated, as though he's aged thirty years overnight. What's more, his skin has become dry, hard and pale, and every day more clumps of hair slough off it.

He turns suddenly and begins to scratch at his back. "It's here, too, right between my shoulder blades, a terrible itch. Scratch it, please! Scratch it!"

Miriam slides her hand underneath his T-shirt and scratches until he calms down. Alan is exhausted, about to fall asleep, when he sticks his tongue out of his mouth in disgust.

"My mouth tastes like disinfectant."

Miriam narrows her eyes but she doesn't stop scratching his back. The tremors she feels mean that his body is reacting to the presence of its invaders.

Swarms of nanites may already have reached his lungs and his stomach. During the next forty-eight hours they will migrate along his spinal column, circulating in the cerebrospinal fluid, until they have taken up their positions on every single damaged nerve fiber. Then the countdown will begin, because, when his own cells are no longer necessary, they will die with dignity. Once the pegs that hold them together have been dismantled, they will devour themselves piece by piece…and then, something else will take their place.

Alan's body tenses – two, three convulsions, followed by an equal number of dry heaves. He clenches at the sheets for a while, until a miserable thread of yellowish saliva dribbles from his lips.

"It will pass, Alan. It will pass."

"When, Mom?"

Miriam knows that what he is going through is called 'apoptosis', or programmed cell death. It happens every day – billions of cells stop functioning, for our own good. In Alan's case, billions of nanites have accelerated the process dramatically. By the time they have finished, they will have altered his biological state.

"Sooner than you think."

★   ★   ★

Alan spends his nights moaning and his days suffering. The light seems to make the nanites more active, and as a result he becomes more listless and short-tempered. Whenever his body isn't tormenting him, he raves, especially at dawn.

"They're eating me alive, one piece at a time. Little microscopic cannibals! Tiny invisible monsters. Make them stop, Mom! Stop them before they devour me."

Miriam knows that water and darkness bring the fever down. She can do nothing else to alleviate the pain that is a part of this less-than-orthodox treatment – nothing but stay at his side for the worst of it, to bear witness to his delirious ranting.

Alan pulls himself up, hauling his legs back towards him. "Last night I was walking, in the street. I think I was in Amsterdam. I was walking and a taxi came. The driver got out and opened the door for me. 'Please,' he said, 'get in. I've been waiting for you.' I didn't understand. I frowned at him, but I started to get in anyway. I couldn't help myself. I had this urge to get into the car and go. Then I heard a song from far away, from the window of a building that faced onto the canal. They were having a party. I stopped what I was doing. I took a step forwards, then I looked back at the taxi driver. I felt like I knew him. He was old but fit, with a fleshy mouth and a face as lined as cracked asphalt. He was wearing a bandanna on his head. I told him I'd changed my mind, that I didn't want to get into the taxi. I wanted to walk. He left, all angry and insulted. He turned up the volume and he was listening to 'Sympathy for the Devil'. It was the same song they were playing at the party. Can you pass me some water, please?"

"We have to do your exercises." Miriam is holding a small saucepan in one hand, while with the other she is mixing something with a large paintbrush.

"Not right now. I can't do it. When I feel a little stronger."

She steps up to the bed and pulls the sheet down to reveal his legs. "Your strength won't come back by itself. Not without the exercises."

"What are the fucking nanites for, then?"

Attached to the headboard of Alan's bed is a large infrared lamp. It's a do-it-yourself treatment recommended by their Chinese friends. Miriam pushes a button, changing the inclination of the bed. "You'll like this, you'll see. We'll get your blood moving and improve your cardiorespiratory capability along with the elasticity of your connective tissues. Are you ready?"

He gives a small nod and the light comes on, illuminating an area on his left thigh. The light remains focused while Alan slowly shifts into an almost upright position.

Miriam lifts the brush, lets the excess drip back into the pan, then begins to paint the warm gel onto his skin. "Do you feel anything?"

Alan shakes his head.

Miriam moves the light to focus on different spots. Although small red spots remain on Alan's skin, the heat causes him no pain, no sensation. He only knows it is happening because he is watching.

"Twenty minutes, then we'll stop."

While she is painting, an unshakeable conviction takes shape in Miriam's mind: the bodies we are given are the result of millions of years' worth of modifications with no meaning, no plan, no purpose if not that of pure and simple adaptation. They are the fruit of a process that has been blind and painfully slow. We have only just discovered it, and it is called evolution. The human body wasn't designed to feel joy (all one has to do is add together one's moments of happiness and then subtract one's moments of suffering to come to that conclusion). It has become the way it is thanks to a set of external conditions. Mother Nature, left to her own devices, would have carried on with her experimentations on the human race and we – incidentally – would have continued to evolve based on random elements in the process of genetic mutation.

Alan puts all of his energy into his next outburst of complaint. "I can't feel anything on my legs, but I'm hot!"

"Wrong. You just think you're hot."

"If what I'm thinking isn't true, does that mean I'm going crazy?"

"It's phantom limb syndrome. Your mind is used to associating certain reactions to certain stimuli."

"What are you doing to me? Phantom limbs, wax all over me, that light on my legs…. Right now, for instance, my bones ache and my skin feels tight." He sticks a finger into his mouth, lifting his lip to show her his upper gums. The teeth wiggle in their sockets as he touches them one by one.

"Leave them alone, Alan. You have to give it time."

"Come out and say it! You want to kill me! It's all part of the plan. You're in league with them! Those fuckers at Globalzon!"

Miriam understands when it's time to beat a retreat. She stops what she's doing and goes out into the kitchen.

"Don't leave me alone!"

"Of course I won't leave you, Alan. I'll be back in a moment."

Miriam listens to him shouting from the other room. "You don't know what I'm going through!"

She hastens back to the bedroom and puts a bottle of water to his lips. "Don't say that. It's not true."

She dries off his chin. Alan can't sit up straight. Only when he's slightly bent over, and the space under his diaphragm is less constricted, does the pain stop. His intestine is almost always empty and, over the course of years – at least according to their Chinese contacts – it will atrophy, together with his liver, reducing the fatty tissue around his stomach.

Miriam wonders what Alan could possibly want at such a time as this, what he's hoping the next three years or ten years will bring him. She doesn't have the courage to ask him anything.

"It is true! You stand there and you look at me. All you do is feel sorry for me!" His body temperature is falling, but his temper is rising.

"I'll say it again. It's not true."

"Liar! You don't know anything!" He's about to cry. He tries to pull himself upright, but there are straps holding him to the bed. "Let me get up!"

"Get up?"

He realizes that what he's said makes no sense. He grabs his guitar from the bedside table and, instead of playing it, the way he often does to calm down, he takes it in two hands, lifts it above his head by the neck and begins to smash it against the wall.

Miriam tries to stop him.

"Stop, Alan. I'm telling the truth. I'm not part of any sort of plan and no one wants to kill you. I know everything your body is going through, every pain you are feeling."

She takes a device out of her purse. At first glance, it appears to be a flashlight, but at the end where the light bulb should be, there's a tiny screen instead. "This is the scanner that I use at work. We use it to identify nanotechnological particles in food. If they're there, I have to put it in a report so that someone else can analyze what sort they are, who they're produced by and whether they are hazardous to people's health."

With a click, Miriam turns on the scanner. She runs it along Alan's body like a metal detector, letting him listen to the beeping sound it makes when it detects the nanites. Then she turns it around and puts it in her son's hand.

"What am I supposed to do with this?"

Miriam guides his hand from her head down to her lap, then all the way down to her knees. The beeping has the same frequency.

This time Alan's tears are different, no longer the angry sobs to which she has become accustomed. In his expression there is something that goes deeper than the changes wrought by the symbiosis with the nanites, something that verges on happiness. What's more, she has shown him that, on the other side of the pain, magic awaits – magic that will allow the nanites to re-establish the connection between his peripheral and central nervous systems and rewrite four million years of natural evolution. Because the nanites will not stop once they have fixed the connections in the damaged Schwann cells. Their work will impact his respiratory, endocrine, digestive and perhaps even reproductive systems.

Miriam's cell phone starts ringing. She lets it continue for twenty seconds before she finds the courage to answer. A week ago, she requested an extension of her leave time from the Personnel Department: another two months at Alan's side. This could be the answer. Once she sees the caller ID, however, she goes into the kitchen and answers in a low voice.

When she goes back into Alan's room, she finds him alert, with two red circles beneath his eyes but, overall, a healthy-looking complexion.

"Who was it?"

"My boss's boss."

"Something important, then?"

"Yes, but first we have to do your exercises."

"What's the plan today?"

Miriam takes four electrodes and puts two on each of Alan's legs, one on his femoral muscle and the other on his calf. "We're going to try 'TENS', transcutaneous electrical nerve stimulation."

"The name says it all."

"This will go deeper than the infrared therapy we were using last month. Still, in your case, you won't feel even a slight vibration, unfortunately. We're doing it to prevent atrophy and spasticity in muscles you can't contract voluntarily."

"That I can't contract voluntarily *yet*."

"Right. Even if you haven't used them for months, we have to be ready."

Miriam presses a button and a current at low-frequency oscillation begins to move through Alan's legs.

"So, what did your boss's boss want?"

The muscles begin to contract on their own. Once, twice, three times. "To fire me."

"Fire you? Why?" Four, five, six times.

"Because people like him aren't human beings. Cecile couldn't do anything about it."

"What are you saying?" Seven, eight, nine, ten times.

"What I'm saying is that, when it comes down to it, industrial patent rights are more important than human rights."

"What now?"

"Now I have to talk to Ivan and you have to think about walking. The rest will sort itself out. Are you hungry?"

She initiates another series of contractions.

"No, and I can't even remember the last time I ate anything."

"Me neither, and I'm not sorry."

When the session ends, Miriam turns off the machine and removes the electrodes. It's then that she notices an almost imperceptible movement of the sheet where it lies over Alan's feet. "Did you do that?"

"Do what?"

"Move your toe."

"I don't know."

"Try again."

"If I can do it, will you make me *bucatini* again?"

"I thought you said you weren't hungry."

"I'm not, but just to celebrate."

CHAPTER SIX

The Last Trip to the Supermarket

The customer authentication system identifies Miriam at the entrance to the supermarket. A cadenced, unaccented voice, like that of a documentary's off-screen narrator, follows her as she makes her way over to the carts, which are lined up in single file.

"Greetings, Mrs. Farchi. Com-pro supermarkets welcome you and wish you a pleasant shopping experience."

Alan limps along behind her, aided by his crutches.

Com-pro has taken the place of every retail grocery store and, in high-technology sectors especially, such as nutraceuticals, it knows no rivals, although some prefer the 'do-it-yourself' option offered by PMCs.

"What would you like for your celebration?" Miriam asked.

Miriam gives cart number seventy-five a push. Its display comes to life and its wheels begin to turn.

"Bucatini, like always."

"Why don't we invite someone? I could make eggplant parmigiana."

It has been a month since the first time that Alan's foot moved. Almost as long as it took him the first time he went from crawling to pulling himself upright onto unsteady knees under Miriam's watchful eye.

"I don't know. I still have *these* things…."

Together they head towards the produce section. They stop, squeeze the eggplants, sniff at the tomatoes, then the zucchini. Both grimace in distaste.

"Listen, Alan. What if we invited Silvia? The rickshaw girl? We could ask her to bring some vegetables from her garden."

Alan grunts, uncertain, self-conscious about being seen in his condition. "Don't forget that she saw you on the day of the accident."

"Fine. She can bring the eggplants, but we still have to buy some things." Mother and son wind their way through the brightly lit aisles of Com-pro, navigating interminable shelves stocked with single-serving yogurt containers, fiber-rich energy bars, low-calorie jams, powdered cheeses, popcorn to 'prepare' in the microwave, and rapidly metabolized pasta. They both feel lost. Around them, the chirping of voices offering advice and recommendations is an indistinct buzz punctuated by jingles and electronically generated animations.

"It's been a long time since we went grocery shopping together. You know, when I was a little girl, there was no self-service. The shopkeeper stood behind his counter and the customers asked him for the things they wanted. People took pride in serving their customers."

"Yeah, there are still delis like that downtown. You pay in cash – no loyalty cards, no coupons or points programs."

"Here they give you those things, but in exchange you have to give them a list of your favorite things." Miriam is sure that the people at Com-pro know all there is to know about her. She imagines electronic folders full of her receipts stored on shadowy company databases. What's more, every time a new special offer pops up on her cart's display, she gets an extremely uncomfortable feeling that those same people could sell her anything and everything until the end of her days.

In the wine section, Miriam picks out two bottles of Morellino di Scansano. A picture of the gently rolling, vineyard-covered hills around Grosseto pops up on a 3D panel the moment Miriam touches them.

"Do you think that will be enough?" Alan asks. "Silvia seems like someone who likes her wine. Grab one more."

"Human relationships counted for something back then," Miriam says. "The shopkeeper knew you, he listened to you and he knew what to recommend. Then came the basic shopping carts – nothing like this brainy thing – and instructions on how to pick things out for ourselves."

"You remember how pathetic those carts were, though, don't you? All they were good for was holding the things you threw into them, and the directions you got from the supermarket were generic, impersonal. But now 'Big Shopper is watching you'!"

Miriam gives cart number seventy-five a little kick. The Artificial Intelligence this model is endowed with makes it the intellectual equivalent of a six-year-old child. "You're right, but the shame is that this cart's Software Agent couldn't care less what my name is."

"It knows your name, though, and a lot more besides."

Under the name Miriam Farchi is a menu from which to access an endless list of information, such as:

types of foodstuffs purchased
list of most/least expensive products
list of quantities
calorie counter
daily/weekly/monthly/yearly shopping
loyalty points/discounts/current promotions

Miriam tries, without success, to turn the display off. "That name doesn't mean anything to this cart. Even if it knows much more than any old-fashioned shopkeeper ever knew about me, it and I have never had a conversation. It's deduced that information from my behavior, through extortion. It's pathetic, to feel like we've been reduced to herds of animals, all needing to buy things, each with our own electronic card, and we can't do a thing about it."

"You're right. It is pathetic. So let's be done with it. What if we invited your friend Ivan?"

As they walk by the candy shelves, the cart's display flashes the offer of the day, calculated using a personalized shopping algorithm. It's for a box of squares of ninety-nine-percent-pure Droste dark chocolate. Beneath the offer appears the cost of the purchases selected so far: €18.55 and, in parentheses, the price of the box of chocolate (€6.99).

"Ivan can't make it. Hold on, let me get some cat food." Miriam reaches across to the pet-food display and grabs a couple of boxes of Kat&Kit.

"Why? What happened to him?"

"He helped me with the nanites and they accused him of violating the Ending Hunger patent rights."

"And this is how you tell me?"

"How should I have told you? He's a hard-headed Russian and he wanted to do it his way. He knew this might happen."

"But why? Didn't you tell me the patent was his?"

"It is and it isn't. The idea for the nanites was his, but the patent was registered by the company he was working for, who filed it under the name of the National Research Foundation."

"They convicted him already? After such a short time?"

"He told me in an email that he'd received a notice to appear in court. Did you know that the courts are being run by Artificial Judicial Intelligence Systems now? They go through the mountains of charges, analyze them and, if need be, convert them into rulings with the same weight as the old ones issued by a Court of First Instance?"

"No, I didn't know that."

"The judges, the human ones, have been shifted into the Courts of Appeal and the Constitutional Court. In the notice, they asked him to appear and confirm their version of events."

"Did they arrest him right away?"

"They took him into custody. There wasn't much anyone could have done about it. A Software Agent paid for by the prosecution had detected the formula on the Silk Road and established violation of copyright. Given that the damages were higher than anything Ivan could ever have paid, he opted to convert his sentence into jail time."

"How long will he be in for?"

"He was given an eight-month sentence. He took the blame, even though I'd asked him to stay out of it."

"One more person I'll have to thank, as soon as I see him."

The cart keeps on spitting out offers, and Miriam and Alan continue to ignore it. "So, what are we going to do about that chocolate? These people know all my habits. They know when I start a diet. They know when I've been disloyal to my favorite brand of tuna because I found a cheaper one. They know on what evenings I'm having friends over to dinner."

"Well, here's what you do. Get that ice-cream cake I like instead, the Party Roll. That'll make their heads spin."

Miriam accepts his proposal and leans over into the ice-cream freezer. There's only one Party Roll left, a layered one manufactured by Mestenè. A moment's hesitation is all it takes and an old woman wearing a pair of leather gloves and fluorescent running-sandals beats her to it. She throws Miriam a wink, but it's clear from the vicious look in her eye that she's ready to fight over her prize and even pull it apart before she'll surrender it.

Miriam doesn't protest and the woman darts away, pushing her cart towards the fast pay line, reserved for electronic payments. Beneath the food-detector, the RFID tags on the items in her cart send up the total amount to debit to the old woman's card. Miriam turns towards her son. "Did you see that woman? Sorry, no Party Roll."

"Forget about it. Hey, at least we know there's someone out there who still cares about getting what she wants. Too bad the shopping carts know how to make sure she's always broke by the time she's due for her next pension check. But, like they say, 'You can't put a price on happiness.'"

"All right, then. Grab whatever's there. A random choice will mess with their calculations."

Miriam grabs a container of Heldigan cherry swirl and signals Alan to head for the registers. Once the wine, cat food and ice cream have been scanned, she pulls out her wallet but, instead of her debit card, she hands the cashier a €50 note. Hardly anyone pays with cash anymore, and the girl demonstrates her annoyance with a huff and a shake of her head. She heads to the office at the back of the store and returns two minutes later, more irritated than before, having managed to scrape together €6.80 in jingling change.

Outside the supermarket, two trucks are parked. One is unloading some pallets, while another is loading up the same amount. Foodstuffs and more foodstuffs, and as the expiration date nears they are thrown away. The emphasis placed on products' freshness, aesthetic appeal and perfection has come back to bite the producers. People won't buy anything anymore if it is not as appealing to the eyes as it is to the tongue. Outside the Com-pro gates, a few carts have been abandoned next to some trash dumpsters.

"See? Every now and again the carts make a mistake. You know, I've seen bottles of shampoo in the shopping bags of bald retirees and women clearly past menopause buying tampons."

"Who knows? Maybe they were shopping for someone else."

"Or maybe they know something we don't."

CHAPTER SEVEN

Serra Spino

On a hill on the far side of the outer Great Ring Road, at the far end of Via Portuense near the suburb of Serra Spino, is the commune where Silvia lives with her friends.

Alan has decided to surprise her by dropping by in person to invite her to dinner. He calls one of those pedal-powered taxis that travel to destinations outside of Rome (he didn't want to call the Pulldogs and risk ruining the surprise) and sits in the back seat, his crutches beside him. Once beyond the muddy expanse that lies just outside the Ring Road, from which rise the outlines of Architectural Compositors and the scaffolding of new construction, he is able to enjoy the autumn colors that line the country road, providing some compensation for the poor condition of the asphalt.

He gets out of the cab, pays the fare, and makes his faltering way up a pathway that runs along a ridge, sheltered overhead by bramble bushes. Two dogs come bounding down the path towards him, a Labrador and a Siberian Shepherd. They have no leashes attached to their studded collars, but they display no hostility. They bark at him in greeting and accompany him until he reaches an iron gate.

Above the entrance hangs the sign from the Bulldog, the famous Amsterdam coffee shop, duly altered to read Pulldogs. In place of the original's yellow dog is the sled dog logo, the symbol under which the team of young people who live here have banded together.

The gate is open. As soon as Alan appears outside of it, he is recognized by the same kid he had glimpsed right before he passed out in front of Globalzon, who motions for him to enter.

On the left side of the courtyard sit six or seven rickshaws. Behind the parking area and running down the slope of the hill he can glimpse the vegetable garden. A few canvases stretched over posts form a crude greenhouse and an artesian well supplies water. A girl wearing a flowered bikini and headphones is watering the plants. Occassionally she makes what looks like part of a dance step. The scent of lemons wafts from her direction. At the center of the courtyard stands an oven constructed of raw bricks, the final element that contributes to the group's diet.

"Hi. I'm looking for Silvia."

There are two men in the courtyard. One is black, muscular and huge of stature, with a friendly face and hair the color of lead. He is preparing to take out one of the rickshaws. The other has Middle Eastern features. He wears a turban and is carrying a toolbox. They don't stop to say 'hello', but both nod in greeting.

The boy congratulates Alan on his recovery, then points towards the farmhouse. "Silvia's inside. I think she's having a bath. She just finished her shift with the rickshaw. You want me to get her for you?"

"No, thanks. If you tell me how to get there, I'll surprise her." Alan and the boy head towards the house together.

"You have to go that way, up the outside steps, and then follow the hallway down to the third door on the right."

Alan is starting up the stairs when he feels a hand on his arm. "Can you manage? You want some help?"

"No, thanks. It'll take me a minute longer by myself, but I can make it."

From the top of the staircase, the property looks exactly like an old farmstead occupied by a group of squatters. A bunch of boys, their skin brown from the sun, appear in the courtyard and start playing ball. A few little girls, their hair in braids, perch on the wall whispering amongst themselves while they watch the boys scuffle.

When Alan steps into the farmhouse, he can hear people chatting and singing. Some graffiti artist, writer or painter from the group has decorated the walls by hand. He follows the hall down to the third door, which is ajar. From outside he can hear the lapping of water and now and then a playful laugh. He peeks inside.

Silvia is washing her hair. She's lying in a curved depression that runs around the rim of what appears to be a luxurious tub of enormous proportions. Alan can only see one side of it, but he imagines it's big enough to hold all of the Pulldogs at once.

The door opens and two wet, naked children come dashing out, giggling. The boy runs straight into Alan's crutches, nearly knocking him to the ground.

"Sorry, I didn't see you. Why are you hiding there?"

The girl backtracks, takes the boy by the hand and pulls him away. "*Ne panimaiesh? On podglyadivaet za ney.*"

"He's spying on her? Why?" asks the boy, innocently puzzled, as they walk away.

The smell of a wood fire fills the room with a sweetish odor. Dry branches are piled on the floor next to a stove that is heating the water.

The racket caused by the children causes Silvia to turn towards the door. Her mohawk cascades down over the left side of her head in a ridiculous fashion. Even while bathing, she's still wearing her anti-smog mask over her mouth. Alan can't stop himself from smiling.

"What a tub. You look really good in it."

"Nice, isn't it? It's one-of-a-kind. We took it from the Hollywood-style bathroom of an abandoned villa on Via Cassia. It belonged to…. Well, who cares? What are you doing here, anyway? Damn! You've made quite the recovery!"

"You see? Miracles do happen."

Silvia emerges from the water, careless of her nakedness. She has two long tattoos, executed purely by hand. They follow the curve of her hips and end on the quadriceps of her thighs, forming a Polynesian design called a *moko*.

"I wanted to thank you for what you did. First for me, and then for my mother. If it hadn't been for you, I'd still be there on the side of the road."

"And you've come all the way out here to tell me that in person? There's such a thing as video-calling, you know."

She starts to get dressed, slipping into a thong, then an old white Run-DMC T-shirt with no bra underneath, over which she fastens the buckles of a pair of denim overalls. Last of all come a pair of purple running shoes with a Velcro closure.

"No, that's not the only reason, even if it would have been reason enough. I'd like to invite you over to my place tomorrow evening. My mother and I are celebrating these." Alan runs his hands down his legs to his knees. "My old legs, back like new."

Silvia gathers a towel from the floor and rubs it vigorously over her mohawk. "How did you manage…so quickly? With all the accidents and injuries we get here every month, I'd love to know your secret."

Alan hesitates for a moment. "If you come tomorrow, I'll tell you."

When the towel falls to the ground, Silvia's mohawk is back to standing four inches tall, spilling open like a fan. "You've got a deal."

"Listen, Silvia, I really don't know how to repay you for what you've done. I've been thinking about it since the day of the accident, and I can't come up with anything that's worth as much as my life." His tone is half-light, half-serious, with a hint of invitation, which Silvia picks up on immediately.

"What about if you take me out some evening?"

"That's an excellent idea. Although, that doesn't seem like much, as expressions of gratitude go. How about if I take you out for the next two years?"

"Careful now. Don't overdo it. I might like it."

"You know that old restaurant on Via Balduina, Rendezvous? Have you heard they've turned it into something new?"

"No. What kind of place is it?"

"I saw an ad for it on a social network. They kept the name but they changed it into something strange. It's actually a perfumery, only that it functions like a club. People go there to try different fragrances and socialize."

"Interesting. At least it's not just another place to eat junk and drink vile crap."

"My thoughts exactly. Speaking of which, my mother wanted me to ask you if tomorrow you could bring some vegetables from your garden. She couldn't find anything decent looking at the supermarket."

Silvia feels flattered. She gives Alan a friendly slap on the shoulder and leads the way back up the hall. "Come on, I'll take you down. Then we can pick them together."

Alan places his feet carefully and every step brings a grimace of pain to his face. He shifts his weight from the right to the left in an unnatural gait, but tries hard to hide it. Inside, he is happy.

CHAPTER EIGHT

The Viaduct

Alan and Silvia sit on the guardrail of the viaduct that connects Via Ostiense to Lungotevere Vittorio Gassman in the Marconi district. Begun twenty years ago but never finished, it has been hanging there ever since, spanning neighborhoods and the river, barricaded and closed to traffic like some kind of road archaeology.

"You're a piece of work, you know. We've been going out for just six months and already you ask me to come and live with you."

Alan gets to his feet and begins to walk over the asphalt, its surface broken by roots and overrun with creepers. Whenever he comes across bits of masonry, he kicks them from one side of the roadway to the other. The long cohabitation of concrete and vegetation has transformed the viaduct into an amorphous structure suspended above the course of the Tiber.

"Much more than that, Silvia. I'm asking you to found a community with me."

"I already have a community."

He strolls up to her with affected nonchalance and buries his nose in the curve of her neck. "This will be very different."

"How so?"

In his eyes there is a strange light, an almost childlike eagerness, seductive and, most of all, contagious. "It's finished, Silvia. The city's lost. We used to go to the cinema, to the theater. Now we just sit and guzzle down streaming videos. Instead of going to a bookstore or to the library, we read on the cloud, and if we have to eat, we don't go to the grocery store. We order on the Internet and wait for home delivery."

"I don't live like that, and neither do the other Pulldogs in Serra Spino.

We all met each other at concerts or at other sorts of places where social activists go."

"That may be true of you, but what about everyone else? The places people used to go to try and meet someone have lost a lot of their customers to online dating sites. Half of people get married because they've filled out forms listing their preferences."

"So? It's not like we've been condemned to live like they do."

"No, we haven't. That's what I'm getting at. Look at this place! We can bring it back to life in any way we choose, compose the pieces in a PMC, make it into the Bridge of the Pulldogs, a magical place." His expression becomes serious. "I mean it. Try to imagine it. I don't want to live my life for some asshole boss who exploits me, not anymore. I don't want to accept a single fucking inhumane condition. The nanites came into my life and my mother's life like a shower of sparks. They freed us from work and the need to eat in a matter of months."

Alan knows that it is a gift that should not be wasted. "I want to build a world, Silvia. Here, in the center of Rome. A place that's different from the one where neither you nor I want to live, the one you and I have never pretended to believe in."

Silvia tilts her head, the better to look at him. The effect that the nanites have had on Alan's body, all of the tiny changes they have gradually wrought, is enormous. When they first started seeing each other, he had barely been able to walk without his crutches. His rehabilitation had gone hand in hand with their walks in the park at Villa Pamphili and the first times they touched, under its spreading umbrella pines. After just a few weeks, his thighs had become as firm as they were in the photos that Miriam had shown her from his days on the high-school basketball team. Strangely, since then, Alan's waist has grown a couple of sizes slimmer, while his chest has broadened.

For six months they walked everywhere. In the evenings they would leave the farmhouse in Serra Spino and, a couple of hours later, they would arrive in the northwest part of the city, walking all the way to Via della Pisana and beyond Casal Bruciato. On the weekends they explored the eastern areas of Rome, cutting from south to north along Via dell'Acqua Bulicante, venturing all the way to the outskirts of Appio Latino and around Via Prenestina. They argued sometimes, when she wanted to stay at the edge of the city, as close as she could be to the countryside, while he always pulled her back towards the center, where they would be more visible, where they would represent an 'anthropological precedent', as he liked to call it.

Alan piles rocks along the side of the ramp that runs down towards the Riva Ostiense embankment, a few yards from the Gazometro, the old natural gas storage tower, a relic from the days of gas lighting. "Instead of working just to enjoy a few minutes of free time, I've decided to enjoy my time directly, without waiting for anyone's permission to take some pre-planned vacation or a holiday dictated to me by the calendar."

Silvia hops down from the guardrail and sniffs at the air around them. "So close to the city center, Alan? They'll clear us out in two seconds flat. People like *us* aren't welcome among people like *them*."

To the left are the clubs and bars where university students go at night. To the right rise the offices of Eni, the oil and gas company, gloomy and inhospitable at any hour of the day. The traffic on Via Ostiense is unabating, noisy, rank. It is a place very different from the one where Silvia is accustomed to living.

"We'll have to set an example. We'll have to be discreet. We'll bring the vegetable gardens to the city. We'll offer our rickshaw services to the people for free. We'll make them love us, so that they'll want us to love them, too."

She listens in silence. Meanwhile, he, seemingly without effort, has already piled up enough stones and chunks of asphalt to form a wall over six feet high.

"They'll hate us, Alan. Or at the very least they won't give a damn. You have to see that."

Silvia kicks at a rock. Alan turns around, smiling wider than before.

"I was like you, too, once. I used to hate everything, and everything hated me. But the nanites have given me a second chance, something that, until last year, would have been inconceivable. I'm talking about how I can walk – and move, and run – without feeling a fraction of the fatigue I felt before, about how I don't have to worry about eating like I did before, but just when I feel like it – once or twice a week, if that, if I train myself to do it."

"You still eat when you're with me." From her rucksack she pulls out a bottle of wine. She lifts her face mask onto her forehead to take a swallow before lowering it right back down over her nose.

"True, but I do it for the pleasure of sharing the experience, not because I'm actually hungry. Think about all the time, money and energy people spend to feed themselves. They work ten hours a day for money. They go to the supermarket to buy food. They take it home, unwrap it, peel it, prepare it, cook it and, finally, they eat it, more out of habit and necessity than for any real pleasure, knowing that there is no alternative to that system."

Alan takes off running towards a no-entry sign fifty yards off. When he gets there he kicks at it with the sole of his shoe. Once he's bent the post it's mounted on, he removes the sign board with a screwdriver, yelling all the while. "And that's not all! Once they've finished eating, they have to clear the table, wash the dishes, throw out the trash and start thinking about their next meal!"

Silvia is almost amused.

"Last of all, most of what they've eaten they'll have to shit out just a few hours later, or else risk getting some sickness or blood poisoning from the same food they need to nourish themselves. It's fucking ingenious!" He carries the sign back to where Silvia is waiting, sits down on the asphalt, takes a can of spray paint from his bag and begins to write.

Silvia goes on drinking the wine without him.

"Oh, right, and if they do get sick, they have to spend more time, more money and more energy buying medicine or going to the doctor for a remedy. And I'm not even adding in the expense of purchasing and maintaining kitchen and bathroom fixtures, all those other products you need to clean them, or the water and energy you need to make them function." Alan puts down the spray can and stands back up, hiding the sign board behind his back.

"Maybe," he continues, "years and years ago, we could have been part of the middle class – before globalization, before the recession, before the Matter Compositors – but not anymore. The only relationship we have with those people," he points towards the buildings and the streets, "is that we haul them around in our rickshaws. We get them to their unavoidable appointments, their meetings and even their dates on time." He holds out the sign board. "Tell me, isn't what I'm proposing worth a try?"

The sign reads:

LAIR OF THE PULLDOGS.

"Are you offering us the nanites? Me and the others from Serra Spino?"

"Yes, if you want them. I don't want to push you to move here if you don't want to. But I'm telling you right now that, as of tomorrow, you'll find me here."

Alan walks away from her, taking the sign with him. At the base of the viaduct's highest tower, he takes off his belt and uses it to secure the sign to his back. Then he turns to face Silvia, beaming. "Believe me, Silvia! I don't want to start the umpteenth useless revolution, with the protests and the demonstrations and blah, blah, blah. Now I know the truth. The root of all evil isn't in our biology, it's in our culture."

He starts to climb, agile, his hands and feet moving like those of a monkey. When he reaches the top, he wedges the sign between the structure's metal shafts so that it will be visible even from far away. From his perch, Alan begins to shout again. "This tiny outpost will be a nucleus for transformation. It will do so much for so many!"

When he reaches the ground, he seems like a new person. "Leave a little for me?" Silvia passes him the bottle and he finishes it in one draught.

"Now, can I give you a welcome-home kiss?"

Silvia raises her mask to accept it. Her top lip is split in two, ensuring she never forgets the close encounter she had with the chassis of a car at the age of eighteen. In the space between her nose and curve of her lip she has inserted a stud of polished basalt, like a beauty mark. He takes in the symmetry of her face, composed of hard and simple lines. Even without makeup, Silvia's lips, their cleft aside, seem sculpted, set like jewels into her skin. When she smiles, her mouth tends to pull up more on the right than on the left. Alan is one of the few who's succeeded in persuading her to let herself be kissed.

In a moment of intimacy, Silvia even confided that, before she met him, she considered sex to be an experience that, while it might be fine for others, wasn't really for her. So much effort spent getting to know someone and then the pain of trying to make them forget you, all for a few rare bursts of joy, stolen orgasms and a feeling of unease, an intimate disquiet. Before Alan, Silvia had been perfectly happy without sex. Then, with him, as if by magic, she had discovered within herself a sensuality that matched her passion for athletics, physical exercise and the open air. She had discovered a partner who found her attractive when she was wearing her running clothes, when she was covered in sweat from pushing her rickshaw and her skin tasted of salt.

Alan takes out his smartphone and taps out a number. "I'm calling my mother to let her know I've found a place to live."

"Wait, first we need to *test* it." Silvia unzips her running suit and starts 'warming up'. She motions for Alan to lie down next to her, on the asphalt. She doesn't like to do it in bed, on the kitchen counter or even standing up in the shower. When she makes love she has to feel free. If she does it the 'normal' way, she feels no joy. When she takes Alan into her, a sort of frenzy blossoms inside of her.

He tries to stroke her face, but she pushes him away. When he tries to raise himself up, she pushes him back down.

"Stop. Let me," she whispers in his ear.

Every time he tries to move so much as a muscle, she stops. When he opens his eyes, her reaction is the same.

"Silvia, I still can't—"

"Shut up. It doesn't matter. You're all I need."

With a sudden shove of his hips, Alan turns her and they roll together, one inside the other. He keeps his eyes closed. When they come to a stop, she's on top again, straddling him, and she keeps moving, surging, thrusting with her hips as her back arches.

Hot tears streak Alan's cheeks.

"Am I hurting you? Do you want me to go more gently?"

"No, don't stop."

"So, you like it?"

He is entering the eye of the storm, love's calmest place. He knows he cannot come but it does not bother him. His lips seek out Silvia's ear to tell her that he loves her, but instead he kisses her, with a liquid heat that does not burn.

CHAPTER NINE

Nanites For All

If she were thirty years younger, Miriam would have been glad to go live on the viaduct with the Pulldogs. It is a relief to know that her son has finally found a place to be – regardless of the fact that an abandoned viaduct isn't exactly the most decent of living situations. Mostly, she is just happy that he has a girlfriend. At the moment, however, she has other problems to deal with. With Alan's unemployment – first because he wasn't able to hold a job, and now because he has developed a categorical aversion to the concept – and her having been fired from the WFP, the money is quickly running out. Her apartment on Via Satrico – three hundred square feet with two balconies – has been on the market for six months. The items of value have all been taken to the consignment shop and her bank accounts, both checking and savings, have been emptied and closed.

"Hello, Mr. Cimali?"

On the other end of the line, the real estate agent has been waiting for just this call.

"I'll accept the offer. I'll be by today to sign the papers."

The Roman real estate market has been at a standstill for years. Over the last twenty weeks, the Inn&House agency has brought an astonishing number of prospective buyers to her home but, for one reason or another (poor credit ratings, temporary job contracts, lack of bank guarantees or simple last-minute changes of heart), all have offered at least forty percent less than what Miriam has been asking for. It had been the inheritance from her father, Cesare Farchi, that had enabled Miriam to buy that apartment, and she wishes she didn't have to sell it off. With the passing weeks, the sale became more and more complicated, until finally the agency itself chose to invest in the property.

Miriam, knowing she is out of options, has decided to stomach a thirty percent cut to her original price.

Alan has told her he wants to found a community. Miriam, meanwhile, intends to vanish from hers. Losing her job has given her a lot of free time, which she spends distributing food to the neighborhood cats. Sometimes she goes as far as Largo Preneste, where there are so many strays that feeding them takes hours.

"Yes, the payment must be in cash."

These days she only goes to the supermarket to buy cat food. It's gotten to the point that the cart has even begun to ignore her, mostly. In the beginning she thought about putting the nanites into their bowls, but then she changed her mind, wanting to save the most precious asset in her possession for a rainy day.

Her plan is an expensive one. There's a payment she intends to make, and then she'll have to watch every euro she spends, especially in the beginning.

"Good. I'll see you later, then."

Miriam's suitcase contains only the bare minimum for her needs. From now on, she'll have to move from one part of Rome to another every time the circumstances demand it. At the PMC, with an investment of five thousand euros, she has composited two hundred doses of nanites, which she hopes will be repaid to her in a different currency – gratitude.

Before leaving the apartment, Miriam takes a final walk through its rooms. She lowers the blinds and draws the curtains in the bedroom. In the bathroom, she checks that none of the faucets are dripping. Finally, she sits down at the kitchen table and looks around at her home one last time.

"Before leaving on a long trip, it is good to sit down. It's a custom of ours, in Russia. We believe it brings luck." Ivan used to tell her that often, before he went anywhere, even just for a weekend away in Umbria or a trip to the seaside. "So that you may sit at the end of your journey," he would finish.

That man, as pig-headed as he is generous, needs her help again. The sale of her apartment on Via Satrico will pay his bail and shorten his 'journey' inside the walls of the Regina Coeli prison. Miriam wonders if, given the ease with which they have ruined him, the same thing is not bound to happen to her. Her money will run out in just a few months, and there are no more Farchis or Cormanis left whose help she can ask. In any case, Miriam does not intend to commit Ivan's error. She will not turn herself in to the authorities like a sacrificial lamb on the altar of copyright.

Her saliva is bitter and has a ropey consistency, but it represents such incredible possibilities. The nanites sustain her body. She doesn't have to flee. All she has to do is disappear.

Miriam rises, takes a package of sunflower seeds and another of shelled walnuts from the cupboard. Then she slips ten cans of cat food into her suitcase.

She leans against the sideboard and taps out an email on her cell phone.

Dear Cecile, I know you're not a bad person. I know you had to execute my dismissal. We've known one another for a long time and there are some things you can believe without knowing them for certain. That's why I'm writing to you, because something is happening. I don't know for sure what the consequences will be, but I can see the signs. After what happened, it seems absurd to me that I did not realize sooner that we are the only living creatures who cook their food before eating it. Stoves, refrigerators, microwaves and every other appliance all appear to me in a new light, one that is macabre and grotesque. I see a chemist's laboratory where our foods are transformed. Instruments of mass decomposition. And I, who have spent my career studying and analyzing food, underestimated the fact that natural substances have been transformed into artificial ones from the dawn of time for this purpose. 'Some discovery,' you might say. Yet, over the past days, I have reflected deeply, and I've come to the conclusion that, in some way, we became 'human' because we do cook our food, because we have a system for producing food and a culinary culture. Now it is this very culture, whose original intent was to free us, that no longer allows us to be free. It does not allow us to choose or to devote time to things that are truly important. So, we might as well stop being this sort of human, stop cooking and eating the food produced by this industrial system and, in doing so, nourish its culture instead. With nanites it could be possible to—

A cell phone rings. Not the one she's writing on, but the other, the one she keeps in her pocket. Miriam stops typing.

"Hello?"

She walks from the kitchen into the hallway. She opens the fuse box and turns off the electricity.

"You're in Piazza Re di Roma? Yes, I can be there in half an hour. I'll see you at the center of the square."

From the other end of the line come questions. Miriam nods and, meanwhile, feels around in her purse for her key ring.

"Yes, I'm well. They hurt a little in the beginning, but I can assure you they work."

Miriam picks up her suitcase, closes and locks the door to her apartment and starts down the stairs, and all the while questions continue to come from the other end of the connection.

"No, it's like I said before, my body gives me nothing to complain about, quite the opposite. I feel like a bird just opening its wings after breaking free from its shell."

PHASE TWO

THE SECOND
LOGICAL MUTATION

'In a big city, if you see a dog going about his business, menacing
no one, fawning on no one, fussing at no one – in fact, behaving
like a good citizen with work to do and no time for nonsense – and
if that dog lacks tag or collar, then you may be sure he hasn't a
neglectful owner, but is wild – and well adjusted.'
Fritz Leiber, *Our Lady of Darkness*

'"No metaphysical mutation takes place," Djerzinski would
write many years later, "without first being announced. The radical
change is preceded by many minor mutations – facilitators whose
historic appearance often goes unnoticed at the time. I consider
myself to have been one such mutation."'
Michel Houellebecq, *The Elementary Particles*

NICOLAS TOMEI

CHAPTER TEN

The Cat Lady

Nicolas reaches into the nutraceuticals compartment, pulls out a tube of curry-flavored Pringles, removes the wrapper, and tucks five layers of starch molecules into his mouth. He always tries to concentrate on the sensory stimuli that the first mouthful offers: the surface of the chip is crisp, in contrast with the softness of its interior. The fragrance of the curry lingers pleasantly, while the flakes of starch dissolve in little explosions of flavor on his palate. While he's chewing, Nicolas waits for the nanomat – a sealed, opaque tank filled with liquid, not unlike an aquarium – to emit a long whistle followed by a short one, the signal that it has finished compositing his coat.

By the time he swallows the chips they have become easy to chew, a bolus of starch, a soft wad of material that liquefies in his throat, so easy to gulp down. Well greased with fats and mixed with saliva, the ball disappears. All it took was a few pumps of his jaw. He recollects that you should chew each bite of food at least twenty times before swallowing. With Pringles, all it takes is five.

For Nicolas, the Cat Lady's response is more than an opportunity. It's something closer to a miracle, fate taking a hand. He checks the display, saves the number he called – although it's possible it belongs to a disposable SIM card that's already gone into the trash – and the text message he received (LARGO PRENESTE 5, 11:45 A.M.: YOU WILL BE CHANGED FOREVER).

It's the third time in a year that Nicolas has decided to change his life, to change his eating habits and move in a different direction, one that breaks his usual pattern of fresh starts put on hold and abandoned soon thereafter. His Medical Agent, a life-saving software that his mother, Olga, recommended he install on his smartphone, has been crystal clear.

"Execute diagnosis." Nicolas lays his palm against his phone's touchscreen. "Risk increment compared to previous scan: +5%."

"How much does that make?"

"Life expectancy: 97.3% elapsed, 2.6% remaining."

"How about in years, months or days?"

"If you do not lose weight as soon as possible, your risk of heart attack will become a certainty."

"Yes, you tell me that every day, but how long?"

"Certainty eliminates probability. It does not indicate a time frame."

Nicolas removes his hand, irritated. "If you can't tell me how many days I have left to live, you're no better than any regular doctor."

The agent, having completed its diagnosis, shows him a graph. If the daily food intake to which Nicolas subjects his body does not drop dramatically, he will have six months left to live. Under the graph is a very long list of harmful ingredients and the relative quantities he has ingested of each. In Nicolas's eyes, it is a death sentence.

At the chronological age of thirty-seven years, his body has been devastated by LDL cholesterol, ACTH hormones, saturated fats, and anything and everything else that might be a hazard to his heart, could destroy his liver and drain his hope. Sadly, the various calorie-regulation apps have had no lasting effect, apart from making him despair yet further of ever being able to stick with anything, even something he wants to do, for more than a few days. Nicolas has lost weight (a little), gained weight (too much), then lost weight again (even less than before) more times than he can count.

He is fed up with the never-ending pendulum of guilt and hunger. He is sick of adjusting the size of his clothing every time his body changes. Perhaps it would be better to get rid of this hindrance of having to think about food and feed oneself five or six times a day, every day. It's too bad

THE ROAMERS • 83

that he reflects on these sorts of things with his mouth full of Pringles.

Nicolas has money, and now he also has a reliable contact, and maybe a chance to ingest a definitive cure. Somehow, within his three hundred plus pounds of flesh, he has never found the courage to take such a step. Every night he comes up with aggressive resolutions about his future eating habits and every day those strict dietary regimes go out the window, whether because of a dinner invitation from his mother or Edora, an unscheduled bite to eat at Rendezvous, or a working lunch he can't get out of. The nanomat's display flashes:

COMPOSITING COMPLETE REMOVE OBJECT

The coat's design is from the spring catalogue of Rakes on Fire, a very trendy shop near King's Cross in London. Nicolas has had to adjust the cut with respect to the original model to accommodate his girth.

He raises the cover of the nanomat and picks the filaments of gel off his new knee-length olive-green trench coat. He pulls it on and looks in the mirror. He turns to view himself in profile, his lifelong weak point.

Looking at himself is not an easy thing to do. Nicolas is not ugly, yet there are days when he forces himself to avoid walking past any sort of reflective surface while getting dressed. This time he scrutinizes his face. He wears a beard, because everyone tells him it slims his features. He has grown his hair to shoulder length for the same reason. He hefts his double chin in his hand, tugs it down, then to one side, then the other, until three evenly spaced beeps make him stop.

"Your adrenaline is rising. Your heart rate has risen above anger level."

"Knowing that doesn't do me much good." Nicolas steps away from the mirror to avoid triggering any further reactions from his Medical Agent and goes back to turn off the nanomat. He covers it with a soft cloth and settles into a padded chair.

The 3D molecular designs revolving in the air above his desk are secondary problems, minor issues in comparison with his meeting with the Cat Lady. The grains of sand in the hourglass projected onto his wall trickle out the time: 11:07. Pietro wasn't going to notice if he was a few hours late turning in his next sample. Still, he couldn't afford to let his father down, not with all the competition from other fragrance designers, all sending their own alluring formulas to Pietro's email address.

Nicolas runs his palm over his smartphone and calls him. "Dad, it's me. I'm going to be in late today."

On the same day that Pietro returned ownership of some of the formulas that Nicolas composed as a minor, he granted his son access to his personal work account. From the emails, Nicolas discovered that Pietro is willing to accept prototypes from anyone. Although being able to compete with others, in his father's view, is a sign of expertise, Nicolas is sure that when he gave him access to those emails it was a stick dressed up as a carrot.

"Fine, but don't forget about the city councilors."

"I'll write to them later, as soon as I get home."

"No, go see them in person. It's better that way."

The previous year, as a slap in the face to his son, Pietro had put an AI Agent specialized in 'computer-to-computer' business in charge of the company. This despite the fact that their suppliers had not been prepared to deal with a meticulous Software Agent with an infallible memory and no sense of humor.

"But I wanted to give you the sample first. Don't you want to try it?"

"That can wait. The councilors can't."

After three months of misunderstandings and complaints, Pietro had reluctantly uninstalled the Software Agent and given his son a desk, taking him out of the fragrance designers' room.

"Fine, I'll go, but right now I have to run."

Nicolas slips his smartphone into his trench coat and gets ready to leave. His home is a minefield of exotic smells and nutraceutical temptations: aromas of Pacific Island vanilla and roasted peanuts, the cloyingly sweet scent of krill wafers and a strong hint of lemon custard.

A spike of ghrelin surges from Nicolas's hypothalamus into his bloodstream, where it translates into hunger. In its thrall, he follows the scent of hazelnuts to the cupboard, where he tears the wrapping from his last package of beta-glucan-enriched Locrackers. He barely chews them at all, indulging his desire for taste gratification, and feels a sense of relief.

The keys to his maxi-scooter are in his hand a moment before his smartphone vibrates in his pocket to a tune by Hot Chip.

Nicolas picks up, annoyed. "Hello?"

"Still there, Nico? I can see you on iMaps, you know."

"Hey, Edora. I'm leaving now."

"The Cat Lady is going to wait until noon, then she'll go to make her usual rounds of the courtyards. With traffic the way it is, you might be late."

"You're underestimating my resources. Don't forget that under all this fat there's the engine of a Lamborghini, pumping away."

"You're not having second thoughts, are you? My friend put his reputation on the line to set up this meeting for you. Don't let me down, understood?"

Edora cares about him. A few years back, when they first met, they'd had sex a couple of times, though not very well, during the filming of a show for Channel 8 TV. At the time, they had both had on-call contracts as studio audience members, but both Nicolas and Edora had gotten bored of spending their days watching faked psychological dramas, the crises of pretend couples, contrived family reconciliations, the testing of artificial personalities, of love compatibilities, jokes on cue and talk shows where they could participate, microphone in hand, judging this or that according to a script. That was why, at the umpteenth predictable dialogue, they had sneaked off to the bathroom together. It had been a quickie that lasted all of one commercial break. By the next episode, she had already found another partner to slip off with, and the next time yet another. They had kept in touch, though, and she had proven herself a good friend on more than one occasion, like when she had found out where to get the nanites. Sex isn't an important enough thing to feel betrayed or offended about.

"I won't. Everything's under control. I just had to make an alteration to my coat."

Nick grabs his Shoei helmet, collects some snacks from a silver bowl that he keeps in the entrance hall for that very purpose, closes the door, activates the Security Agent, and whistles for the elevator.

"You downloaded a new jacket to go to the meeting?"

"It's not just some jacket. It's a fantastic trench coat by Rakes on Fire."

The elevator discharges Nico into the garage beneath his building, into the sixteen-by-nine-foot vault where he keeps the Beast while it's charging: an enormous Lamborghini Vestax II maxi-scooter, shiny and streamlined, assembled with the help of his colleagues from the aroma-bar's fragrance-design lab, Sudhir and Mirna. It had taken nearly one hundred uninterrupted sessions and five days of data-processing distributed over three different PMCs to composite the four hundred pounds of mechanical matter that make up the maxi-scooter – a lot less time than he'd have had to wait to purchase one from a dealership.

"Well, then, you'll make a great impression, and tonight we'll head over to Freni & Frizioni and raise some hell. What do you say?"

When Nico climbs onto the Beast he feels lighter. His cheeks press against the soft protective interior of his helmet. Edora is about to hang up, unable to compete with the harsh bursts of sound from the throttle.

"Maybe. I'll call you later, when it's done."

"Come on, Nic. This is finally it, I know it."

With a powerful acceleration, Nicolas is out of the garage and entering the flow of traffic that runs along the Tiber on Ripa Grande. The press of vehicles in front of him forces him to slalom between lanes, braking suddenly only to open the throttle wide again a moment later, zigzagging along at fifty miles per hour. This swerving through traffic gives him an abstract sense of power, that feeling that comes from avoiding death by a fraction.

Then, the brakes snap into action. The Beast screeches from fifty to fifteen mph, leaving a trail of rubber on the road behind it. At the intersection with Via Porta Portese, speed is electronically reduced by a Traffic Safety Agent that forces vehicles to within the permitted speed limits regardless of their drivers' actions, through a combination of sensors, GPS, and the black box that every vehicle is obliged to carry if it is registered for use on the road, even if it was composited at the PMC.

While he's waiting for the green light at the first info-signal, Nicolas looks down at his jacket. It's tight across his gut. The body stores up calories today to use tomorrow. But tomorrow is never any different from today and the layers of fat accumulate. Adipose tissue is like money you earn. You put it in the bank and spend it when you need it. Nicolas has an excess of both of those things. Food and money. Money and food. He hates them both.

Above Lungo Tevere Aventino float advertising clouds in garish hues. They change shape and color, forming the images of different company logos and trademarks every thirty seconds. It began when the companies offered to shoulder the burden of managing the city's UVA-protection system. In exchange, the municipal government gave them a license to use slices of the sky for advertising. On the outskirts of the city, they get air-traffic corridors that have fallen into disuse. In the center, volumetric lots are assigned through public auction.

Nicolas accelerates and shoots beneath a series of promotional rainbows, a weather system of advertisements that changes to suit the seasons, unavoidable for anyone looking skywards. Just Eat Us – a galaxy of artificial cirrus clouds drip icing and pour down rivers of sugary sauces in the sky near Nicolas's house, above the ancient Apostolic Hospice of San Michele, now converted into a convention center. Between the Tiber Island and the Palatine Hill hovers the compact, squared-off nimbus cloud of The Taste Station. It's the newest model of taste distributor, also known as a 'flavenhancer', capable of intensifying flavors, varying the sequence with which they are perceived from the moment of the first bite, delaying or accelerating the time it takes for the taste of any given nutraceutical to be released onto the taste buds.

The sky above Rome is visible only through a series of corporate filters.

Nicolas sniffs at the air, dilating his nostrils, and the aroma of grilled meat tickles his olfactory receptors. The old embankment beneath the Orange Garden has been dug out, and from deep within it rise tendrils of tempting scents from one of their competitor aroma-bars, Not by Nose Alone.

Twitching his lips to the side, Nicolas uses his teeth to grasp the beverage tube that runs down from a reservoir built into the lining of his helmet. He sucks down a long swallow of coffee-flavored Black Bird sports drink. Every time he goes over a pothole or a bump in the asphalt, the weight of his body taxes the Beast's suspension and makes the joints of the chassis creak. When

Edora is riding hunkered down on the seat behind him, the strain is nearly enough to throw a rod or blow out a tire. Red info-signals are a relief not only for the scooter, but also for Nicolas, who can slip a hand into the deep pocket of his trench coat and pull out a Cocorich to nibble on.

Surrounding Nicolas, in cars and on motorcycles, on scooters and in rickshaws, beneath the gaping eyes of tourists being funneled along special audiovisually enhanced routes, many others are doing the same.

Nicolas shifts into first as he turns onto the road that runs by the Circus Maximus and he feels his despair disappear, banished by the Beast's acceleration as it rears up onto its back wheel on the rise that runs parallel to the ancient chariot track, then over the top of the hill, past the public rose garden, until it comes to the next speed-check, at the intersection with Viale Aventino.

Nicolas raises his mirrored visor and stretches out his legs, first the right and then the left. Already, in the generation preceding his own, a third of Romans between the ages of twenty and seventy-four found they had become overweight. It had been a phenomenal increase, distributed, as it was, across every age group, regardless of race or sex. The development of nutraceuticals had led to the replacement of everything edible that the food industry had to offer, just as PMCs had rewritten the processes of purchase and consumption for nearly all types of goods and property.

Nicolas was the archetype of a transformation whose ripples had spread quickly, easily and systematically. A succession of avian flu epidemics had convinced people to cut all poultry from their diets, while a series of scares about pathogens moving freely between continents had led to embargoes on entire national economies, forcing producers to slaughter their cattle and pigs *en masse*, to the point that the destruction of livestock no longer even made the news. The TV news programs had been the first to spread the rumor, later statistically confirmed as fact, that while every sector of industry had suffered as a result of the advent of PMCs and the household nanomats that followed, the traditional food industry had been entirely wiped out by it. Dependence on a supply of perishable resources, the ludicrous global procurement system and the ever-present health-security fears had all decreed its collapse. Cooking had become a luxury, a quaint experience for those nostalgic for culinary curiosities or for those oddball enthusiasts who coveted the sort of specialties formerly sold in delicatessens. Along with the disappearance of food 'grown or raised on the land' came another phenomenon. The hundreds of restaurants that had once filled Rome's historical center, the last bastion of the concept of traditional 'Made in Italy' quality, began to close at an astonishing rate, either converted into or replaced by the sort of low-calorie socialization services offered by the new aroma-bars which, with their fresh and volatile enticements, had staunched the hemorrhage of customers. Young people made up the majority of their clientèle, flocking around the 'scent diffusers' just as they had once gathered around the tables of bars and restaurants.

At the next red info-signal, in Piazza Numa Pompilio, Nicolas rummages inside the belt of his trench coat, where there is a soft, layered band that he has filled with jelly candies. That particular info-signal lasts long enough to allow him to finish his Black Bird drink and stock up on calories. When the signal turns green, Nicolas is replete. He takes a deep breath, unleashes the power of the Beast and shoots onto Via Druso, leaning perfectly into the curves on Via dell'Amba Aradam, imagining his meeting with the Cat Lady. Edora's friend described her as an older woman dressed in crinoline with a bemused air and an obsession with felines. He even hinted that the only reason she even bothered with illegal dealing was to provide for the cats, saying she continued to bring them cans of food even after she had already fed them nanites.

Some routines become a reflex. Nicolas's reflex has turned into a vice, and now he intends to lose it. He tries to imagine the look on Pietro's face if he could see him without those extra hundred pounds of flesh. What would Olga say if his life expectancy rose back to being over eighty years?

Nicolas does not know how or why he lost control of his food intake. While some have 'proven' that eating habits are physiological in nature, others have shown that they are psychological. Nicolas knows only that he needs to stop deriving so much pleasure from simple, basic things like sugar, fats and salt. Numerous times he has tried to consider nutraceuticals from a different perspective and, while he acknowledges that celluchips offer the benefit of sustaining him, he simultaneously blames them for having made him put on more weight. Sheets of pressed meats and gelato-gels may stave off feelings of hunger, but they have also made him a slave to a repetitive ritual that is killing him.

He knows it's true that, in the end, it's all hunger that exists in the *mind*, not the gut.

As the road narrows, curving by the Church of Santa Croce in Gerusalemme and under the arches of Porta Maggiore, Nicolas struggles to recall the last time he was truly hungry. Accelerating so hard he lays rubber on the pavement, he realizes he's hard put to remember.

After three green info-signals in a row, he finally parks his 'Lambo' at an angle at the entrance to Largo Preneste, in the shadow of the expressway. The clock on the scooter reads 11:44.

Nicolas takes a short stroll around to get his bearings. Whenever he walks, he feels like he's dragging along a suitcase without wheels. He finds a spot behind a low wall and sits down on a tree stump to wait for the Cat Lady to make her appearance.

Ten minutes later, there's still no sign of her.

The façade of the storefront across the way, Xin Xin – Massages and Relaxation, is plastered with life-sized decals showing moving images of five different girls in provocative poses. They smile at him from the first, second and third floors. Every so often they point down in the direction of the entrance or up towards the telephone number and email address. On the first floor, one of the decals has gone missing. Someone has taken it home.

To kill some time, Nicolas lights an anise cigarette. It's not long before a *smartdust* cloud appears over his head. It's composed of swarms of weather drones, temporarily diverted from their task of reading the logs of merchandise in transit on the expressway above to take samples of the air above him, in search of harmful gas compounds. They flash in the air around him, giving off a pale greenish light to signal a negative reading, meaning that the tendrils of smoke examined by their pollution sensors have been approved by the local control center.

Nicolas knows what is permitted and what is not. Every scent he's wearing he has composed or modified himself. Among the Rendezvous's clientèle are two city commissioners. One, Fulvio de Caro, is the Commissioner for the Environment, while the other, Giovanna Ferri, is the Commissioner for City Planning. They're expecting him today. Municipal levels of olfactory pollution are monitored using warning thresholds that Nicolas himself draws up in collaboration with the commissioners every six months.

His father has always had a good nose for business, so, ever since he was a child, Nicolas has been trained for the impending olfactory revolution. When he was a boy, his father would come in while he was sleeping and hold a cloth saturated with some specific scent blend near his nose. Nicolas would become restless, his breathing would change but, if the scent were not too strong, he would not wake up. Every night, Pietro would subject his son's nasal mucosa to a carousel of odors and, every night, Nicolas's reaction diminished. There had come a time when he'd ceased reacting at all. At the age of eleven, Nicolas had already known most of the fragrances on the market and, to Pietro's mind, that constituted an excellent competitive advantage, not to mention a secure investment in the future of the Tomei family.

After a while, Nicolas begins to notice some suspicious goings on at the corner of the street. A little brunette, about five foot three inches, with tan skin and an explosion of curls radiating from her head, impossible to ignore, is pacing back and forth between the intersection's two info-signals. Meanwhile, on the other side of the roadway, a Subaru carrying a couple of old men with large noses and thick glasses is already on its fourth pass in front of the Cat Lady's building. In the end, a siren puts an end to the mystery. A sedan shoots out from behind the building, hits its brakes hard and comes to a stop next to a newsstand, blocking traffic.

Just then, an older woman with a wide-brimmed hat lowered over her eyes comes out of the front door, escorted by two men in plain clothes, although there's not much else plain looking about them. One has a face that's scarred and pitted, and the other has a jaw like a comic book superhero. They shove her summarily into the car and signal for the waiting driver to pull out.

The early afternoon sun casts the slanted silhouette of the expressway onto the buildings lining Largo Preneste. Immediately, Nicolas makes a phone call.

"Edora, she got picked up."

"Nico! What's going on? What are you talking about?"

"The Cat Lady. She just got arrested."

"Oh, Nico. Did she really? Damn it! I'm so sorry."

He wants to curse, shout, maybe smash a window or set the newsstand on fire.

Instead he shoves a hand deep into his trench coat pocket.

"What are you going to do now?"

Nicolas, chewing, gets back onto the Beast and starts the engine. "What do you think?"

"Come on, you know it's not good for you. You need company. I'll be outside your house in twenty minutes, all right?"

"Okay. But don't come empty-handed. I feel like shit."

On the way back home, Nicolas's thoughts swerve back and forth just like the Beast as he zigzags through the traffic on Via Druso. His tendons are so tight and strained from squeezing the brake lever and opening the throttle that he has to concentrate just to orchestrate a pass, skidding as he veers to avoid a car.

The Beast growls and spits heat from the exhaust. Behind the visor of his helmet, Nicolas curses through gritted teeth. One risky move, one angry huff too many, and with a pop a button flies off his trench coat, leaving it open over his gut. Nicolas hunkers down over the gas tank, draping his body atop it, overcome with disgust for this envelope of flesh that *always* needs to eat, so lacking in stamina that he has to lie down every eight hours, has to take it to the toilet even more frequently so that it won't embarrass him any more than it already does.

He works his way hard up through the gears until he's redlining downhill in fifth, then downshifts savagely at the thought that, over the course of thousands of years, nature and human evolution haven't been able to come up with anything better than this. Something that requires less care and attention. Something more lasting and not so easily fouled up.

The Beast devours one yellow signal and then another, the second by a hairsbreadth, before he's forced to slow down by the red info-signal across from the packed-earth track of the Circus Maximus. Next to him, at the front of the crowd of vehicles, is a horde of rickshaw drivers, slender and athletic, entertaining their passengers with tales of the Ancient Romans while they wait for the light to change. Some recite bits of real history they've learned by heart, others recount half-invented legends, while yet others rely on an artificial tour guide that suggests to them what they should say as they glide by the various sights. They are all slim, but it's not because they're poor. It's clear that they burn their calories. Nicolas, on the other hand, burns rage and hydrogen.

He is crossing Ponte Palatino bridge when a Bonfood delivery van begins to swerve. Its driver is doubly distracted, carrying on a shouted phone conversation while compositing something in his dashboard nanomat. In a split second the van has jumped the curb that divides the lanes and come crashing down onto a charcoal-colored SUV, which flips over onto the rickshaw directly in front of Nicolas. He is looking at the silhouette of the Dannon obelisk when it happens, a hundred and thirty feet of granite layers assembled on-site in Piazza in Piscinula by an array of six PMCs. Over the sides flow a random sequence of the brand names distributed by the food-industry colossus.

What he is actually thinking about is the sweet-and-sour lemon and lychee soup he gets from Kim's food truck, and how he could stop there for a restorative snack.

Nicolas swerves, bears down hard on the brakes and crashes against the bridge's iron guardrail. He and the Beast separate in mid-air, its front fork jammed into the railing's metal grille while Nicolas is thrown ten yards forward. For five seconds he remains on the ground, immobile. Then he raises his visor and puts a hand to his bloodied mouth. His beverage tube has cut his lip open. He shifts his gaze to take in the scene. A man is lying in the road, moaning, arms stained red. Next to him, the wrecked frame of a rickshaw, its torn soft-top emblazoned with the image of a sleddog. Nicolas has seen many similar accidents, but always from a more detached perspective.

"Don't move, Nico. I was behind you. The way you flew...."
Edora is kneeling over him, her voice shaking with fear. Her fleshy fingers
intertwine with his.

"What am I going to do now, Edora? What rotten luck. They got her.
Today of all days, Edora. *They got her*, do you understand?"

She removes his helmet gently. "It doesn't matter. We'll find
another solution.

"Come on, let's go home. Can you walk?"

She helps him to roll over onto his side and then to sit up. "Yeah, I
can walk."

An ambulance siren is approaching from FateBeneFratelli Hospital and
behind it, softer still, a police siren.

"It's just my luck. This would have to happen, too. My hip hurts, and
my knee, and my elbow, here, but no ambulance, please. Tomorrow I have
to deliver a smartfume and I can't fuck that up."

Nicolas hands Edora his smartphone. "Go on the website for the Road
Maintenance Authority and call a tow truck to take my scooter back to
my place. While you're waiting, have someone help you get it up and
parked. The police will never know it was involved. I have nothing to do
with this."

He struggles to his feet and leans against Edora's car. "Look at that. I just
composited it this morning."

He sticks a finger into the rip in the graphene trench coat where an
abrasion runs down the back of the right sleeve. Whatever else, it had saved
his life.

CHAPTER ELEVEN

Smartfume

"It's still there, the bastard."

Nicolas stretches, feeling tired. He's spent at least fifteen hours lying down but, now that he's up, he feels like it's been only one. He's standing in the kitchen in front of the nanomat he's just switched on, tearing huge bites out of a slice of tomato and mozzarella pizza, snatched cold from the fridge. Before getting down to work, he always follows the same ritual: he checks fragrance sales while having a snack.

On the nanomat's display he can see the ranking of the most popular scents at Rendezvous.

In first place there is still *that* blend of tropical orange and mango with added erotic stimulants, plus omega-3 and restorative oils. It's called Shandy, and for months people have been downloading it at rates bordering on the obsessive. Nearly once every twenty seconds the formula moves from its virtual state, recorded on the aroma-bar's servers, into the form of a liquid essence on a nanomat tray, only to finally become volatile, nebulized high and low. In the daytime, it's most often used in cafés, but in the evening it's used in dance clubs – a scent to 'breathe new life into anyone's day or night', as Pietro would put it, and he's the one who has the final say when it comes to advertising slogans.

It had been Pietro who had purchased the fragrance that was to form the base for Shandy at the Manaus Vapor Exchange, from a consortium of regional fragrance designers. Then, as usual, he had asked Nicolas to alter it.

"Now I'll show you."

Nicolas composes best with his mouth full, when he's not thinking about the fact that the value of his fragrances, sold in the most fashionable aroma-bars in Beijing, Mumbai and Buenos Aires, is an amount that contains a whole lot of zeros. Then again, it's been a year – fourteen months, to be exact – since the last time one of his original smartfumes reached the top of the charts. That one was called Allublù, a tablet that, once swallowed, ensured the user a week's worth of perfumed perspiration. No other composition of his has achieved such results since. That is because Pietro – who had always been the one who managed the practical aspects of the business, like copyright registrations, relationships with suppliers, licensing and marketing campaigns, who had always left the task of creating fragrances and surprising the market from one quarter to the next to Nicolas, the designer of the family – had decided instead to move him to the second floor, to the business office. His salary has doubled, but his time for 'creation' has been limited to the nocturnal hours.

Nicolas cracks his knuckles and opens up Nanocad, the molecular modeling software.

It's the fault of the nanites that he's lost a day of work and that he didn't stop by to see the city commissioners.

It's the fault of the Cat Lady that he destroyed the Beast and he hasn't been 'changed forever'.

Luckily for him, at least the only reminder of the previous day's accident is a sharp pain in his elbow, nothing more. Actually, that's not true. His hands feel hot, like they're burning, even though he hasn't even begun modeling.

"It's time for me to take you out of circulation, Shandy. It's high time I took back my spot at number one."

The flourishing fragrance market has undergone profound changes since the introduction of PMCs, and the advent of nanomats has brought about even greater ones. The old concept of fragrances still exists, although it is slowly giving way to the new and intriguing idea of the smartfume: a dynamic essence capable of making a given scent match up with a customer's desires as they move from situation to situation.

Another bite of pizza, to restore his concentration. What a ridiculous sound that name had, Shandy.

While composing aromas comes naturally to him, finding a *nom de guerre* for every fragrance, so that it can jockey for place amongst a thousand other essences, is a task that Nicolas gladly leaves to Pietro. He is adroit when he's tinkering with molecules. He can call to mind an object's formula and visualize it as if it were a scale model, in perfect detail. When it comes to using words, however, for him it's always trouble. While formulas and phrases might both have complex structures, Nicolas prefers to focus on the former and avoid the latter – so much so, in fact, that his colleagues at Rendezvous affectionately call him the 'poet of perfume', a soubriquet that, even now that he is over his initial embarrassment, still often irritates him. Nicolas considers himself nothing more nor less than a *fragrance designer*. He sees nothing poetic in his creations. They are atomically stable structures, governed by the principles of physics. His profession differs from that of the chemist, in that the latter mixes molecules in a solution until they collide with each other at random, while the fragrance designer arranges atoms and bonds them in a linear sequence – in orientations and patterns that may vary but, once the product is complete, are entirely one-of-a-kind.

Eliminating the randomness of molecular movement prevents the occurrence of unwanted reactions, of an instability that would increase as the dimensions of the composition grew. There is little of the artistic or sensual in such a process, although people often start off their evenings in the aroma-bar only to leave for more secluded settings, abandoning themselves to an arousing fragrance and giving in to the power of an aroma. Nonetheless, Nicolas knows that the best scents, the most successful ones, the only ones that are still remembered years down the road, are the ones that are capable of bringing people together at a chemical level, whose molecules can be partaken of and experienced together, each through his or her own nasal passages but which create a mutual emotion, a 'shared sensory experience' – or, in this case, it might make more sense to call it a '*scentsory* experience'.

While it is true that scents can be blended together to create millions of subtle variations, it is likewise a fact that they cannot be broken down into a few component parts. It is this aspect of their nature, this olfactory irreducibility, that annoys Nicolas more than anything else. Sounds and colors can be represented mathematically, visualized on a computer and consequently recreated, based on their unique frequencies, using numbers.

Scents, however, do not obey such rules. One must sniff at their roots, understand their essence and look them up, one by one, in a hypothetical olfactory dictionary – as if such a thing could ever exist. For Nicolas, that dictionary is his nose.

He selects a molecule of methyl phenylacetate, with an aftertaste of honey. He lowers his eyelids and flares his nostrils.

"Too delicate, too volatile."

The effect would last the space of a breath, the wind would dissolve it at the first gust.

He adds an antioxidant, a few molecules of oxalic acid derived from Amazonian cacao, and a common carotenoid, to make it all seem more familiar. The molecules bond without any difficulties. The stream of scent that emanates from the nanomat diffuses, striking Nicolas with a fragrant jet of air. He wrinkles his nose nonetheless. So far he has focused on delicacy, but not on structure, and he hasn't even begun to work on persistence or, most importantly, depth.

Moving very slowly, he leans in over the tray and inhales once more, as deeply as he is able. He moves his nose from one side to the other, noting that the scent leaves a trail that is too heavy, a distinct and static tone.

"Either something's missing or there's one component too many."

A fragrance, when it is at its most honest and truthful, serves to create a distortion, a sort of sensory mantle that convinces those who are wearing it of their own appeal.

"It's the honey...too bold a reference. Too cloying."

Nicolas selects that particular molecule and, with a gesture of his hand, deletes it. The filament evaporates, one atom at a time. A few nuclei rebound against the others, eventually reattaching themselves to the main filament, forming minuscule crystals like jewels with a perfect mineral structure. Others come together, seizing hold of free valencies.

His smartphone lights up and launches into a thunderous playback of The Who's 'Baba O'Reilly'. Nicolas knows who that ringtone is for and he has no intention of stopping in the midst of his whirl of creative inspiration to answer, but insistence is typical of Edora. If he doesn't answer, within two minutes she'll send him a message, then another, and then another, in a nerve-grating pattern hazardous to their very friendship.

She weighs him down, figuratively as well as literally. Nicolas taps on the display. "I'm working, dear."

"I figured as much, dear. I'm just worried about yesterday. Tell me, how are you?"

He goes into the kitchen, takes another piece of pizza from the fridge and puts it into the microwave.

"I'm fine. Listen, since I've got you on the phone, why don't you give me a ride to Rendezvous? I haven't even had a chance to download the formula for the replacement part I need to repair the Beast."

"I can't do it, Nico. I have to be at the office in an hour. I can give you a lift home, though, after work. What do you say?"

Fifteen seconds and the slice of pizza is in Nico's mouth. "Okay. I'll see what I can do about getting a ride."

"Make sure you take it easy today. Tonight, I'll make sure you have a good time." Nicolas hangs up and goes back to the nanomat to finish up his formula. His hands feel like they're on fire. He goes into the kitchen, opens up the freezer door, sticks his arms in up to the elbow and stays like that, his belly propped against the frosty edge of the freezer compartment.

A minute later, his hands feel cool. He rests his fingertips against his forehead and, massaging his temples, walks back into the living room. He selects the 50ml option on the nanomat's display and presses the start button.

"You've forgotten to close the freezer door."

"I'll get it in a minute."

A very fine powder forms on the tray, condenses, then liquefies. The final product drips, a drop at a time, into a phial bearing the label

PROPERTY OF RENDEZVOUS AROMA DESIGN – ALL RIGHTS RESERVED PLEASE DO NOT SMELL

"Two more minutes and I will be forced to take action to prevent damage to the automatic defrosting system."

"Fine, you close it then."

Nicolas grabs his smartphone and heads into the bathroom. He sits down on the toilet seat, which he never raises – a simple question of laziness. In his mind he is already preparing to resign himself to this humiliation, which he views as a punishment: the legacy of our animal state, offensive and somehow dehumanizing. While he searches for a taxi service on the Internet, a sudden sharp pain forces him to double up over his own gut. Could it be the first sign of the danger predicted by his Medical Agent?

"Fuck you. You told me six months."

"I have nothing to do with it. It is all your doing."

"Will you please tell me if this is a false alarm?"

"If it is a false alarm, why should I?"

Nicolas engages the app's force quit option.

"It is no use. You disabled that option yourself. Your life depends on it. Remember?"

Nicolas hurls the smartphone against the wall. Not too hard, but quickly enough to stop the app's anger-control reaction from kicking in.

Ridding oneself of what one has eaten is a horrible thing. During these moments of forced evacuation, Nicolas would like to eliminate the anus from the system of natural excretions. While he's at it, he'd like to do away with his mouth as well. Looking out of the window, towards the Orange Garden on the Aventine Hill, he hopes that, one day, nanites will be able to help do something about that, too. If only he had arrived earlier for his meeting with the Cat Lady. If only he could find another dealer to supply him with the nanites. He has always longed to find the 'perfect food', one that satiates without making you have to leave organic waste lying around like some animal, and each creature's excrement – Piazza de' Mercanti is a sad showcase of this – bears its own pitilessly distinctive biological mark.

That gets Nicolas thinking about dogs and, in particular, of the sled-dog symbol printed on the soft-top of the rickshaw that was destroyed in the accident. The word on the logo had read Pulldogs. He picks up his smartphone and googles it. The first search result is a website with a toll-free phone number.

"Hello? Rickshaw service?"

THE ROAMERS • 101

As soon as he stands up, the toilet sanitizes itself.

"I have to go to Prati. I'm in Trastevere, in Piazza de' Mercanti."

He goes back to the bedroom, opens the closet and selects a serious-looking suit, one befitting a businessman, and a tie with diagonal gray and light-blue stripes.

"All right. I'll be downstairs waiting in ten minutes."

He sets the smartphone on the bed and gets dressed. It's a shame that, once he's wearing the suit, it looks ridiculous on him, bordering on clownish. He looks like Olga dressed him.

He goes into the kitchen. When he feels hungry – the kind of hunger that spills over into anxiety, nervousness and fury – it's almost enough to make him cry. In the end, he unwraps a packet of eight individually packaged croissants and is filled with an immediate joy as the uncontrollable pleasure of caloric intake enfolds him. His happiness, however, soon bursts its banks, rising along with his cholecystokinin, surpassing the sensation of satiety to become guilt, the desire to douse that cloying sweetness with something to drink. On the refrigerator, Nicolas has hung a wretched but useless deterrent: his ugliest photos, the ones where his double chin shames him and his gut bulges over the waistband of his pants in rolls a hand's breadth thick.

When he can't bear to look at himself any longer, he closes his eyes, opens his mouth and gulps down a pint of lemon Sprint-Up.

The Pulldogs kid is lean, angular even, underneath a stretched-out T-shirt on which random messages appear and disappear at regular intervals. From the base of a head smooth as a watermelon a single tail of blondish hair splits into two thin braids that hang down to his shoulders. Another strip of hair runs straight down from below his lower lip to his Adam's apple.

"Hang on, I'll help you up."

Not sure whether the boy's offer is prompted by his limping from the previous day's accident or the weight he carries around with him, Nicolas refuses the extended hand and gets up onto the seat without help. As soon as he settles into place, the rickshaw's suspension begins to creak.

The boy shoves hard against the bar, putting his back into it to get started, and they begin to move forward.

"Where am I taking you, exactly?"

"To Rendezvous, in Piazza dei Quiriti, in Prati."

The driver turns on iMaps, raises the soft-top and sets off through the alleyways of Trastevere, around the Church of Santa Cecilia, heading towards Piazza in Piscinula.

"Rendezvous, eh? I've heard of it. Got a hot date?"

Nicolas is in a foul mood, which the fiasco with the Cat Lady, his scooter accident, the smartfume he has to deliver to Pietro and the kid's insolent 'helping' hand have done nothing to improve. He is not in the mood to chat.

"Actually, I work there. I'm a composer."

"A composer? That's really cool. So, you work with, like, 3D formulas and atomic models and shit. Light-molecular industry and DIY, right?"

Despite the compliments, the kid's tone is patronizing. As if pulling people around in a rickshaw is such heavy work, as if doing it makes him some kind of tough guy.

"No. I compose fragrances."

"Oh, sorry. I didn't mean to—"

"Don't worry about it. Listen, do you mind if I have a little nap until we get there?"

"No worries, brother. Make yourself at home."

The rickshaw's seat is comfortable, made of a shock-absorbing material capable of making even the bumps from the worst of Rome's potholes disappear. Besides – Nicolas thinks to himself as he tries to find the best position for leaning against the headrest – as far as the view is concerned, it's not like he's going to miss anything. Around him, the economy is promoting itself by means of trademarks, its pride and joy, and it is doing so with imagination and persistence. There are brands that are born and die in the space of a few weeks, disposable apps whose names come and go too quickly to be remembered. The blueprints for the products people request on the web are sent to their household nanomats to be fabricated with different personalizations every time. Purpose-made medical implants, customized facial jewelry and sports shoes molded to fit someone's bony or calloused feet – all are products that can be composited in just a few clicks.

People continue to have jobs because, just as manual laborers once disappeared to be superseded by office workers, now those office workers have in turn been replaced by molecular compositing designers and specialists. While nanomats have rendered obsolete most of humanity's manual tools – and the knowledge of how to use them – there is still a need for people who know how to make the nanomats work. The Distributed Intelligence Systems have not yet become Creative Intelligence Systems.

The first drops of rain begin to patter on the soft-top the moment Nicolas closes his eyes. Rome's spring cloudbursts are sudden downpours of lukewarm water falling from a yellowish sky that verges on ochre.

Near St. Peter's, the Pulldogs kid pulls over to the curb and turns to face Nicolas. He checks to make sure that his passenger's eyes are moving rapidly beneath their lids, then slips a hand into Nicolas's pocket and pulls out the vial of smartfume. He shakes it, opens it to sniff the contents.

"Sorry, composer, but who knows how many more smartfumes you have in that head of yours?" he whispers with satisfaction.

Then he resumes pushing the rickshaw towards its destination.

CHAPTER TWELVE

Rendezvous

Pietro Tomei is waiting for Nicolas by the club's patio, arms crossed, barely masking his annoyance at seeing his son arrive in a piece of junk pulled by some lowlife. Nicolas wakes, sits up, and hurries to step out of the rickshaw, pulling out his wallet.

"How much do I owe you?"

The kid shrugs and spreads his hands wide, palms up. "Not a thing. The Pulldogs pull for free inside the Ring Road. Didn't you see that on the website?"

"Oh, no, sorry. Haste and habit, I suppose."

"Haste and habit are ugly beasts, brother. Call us if you need us again."

Nicolas is already standing on the asphalt. Pietro greets him outside the club, as he does every day.

"Does your arm still hurt? Where are your bandages?"

The previous evening, to justify not having met with the commissioners, Nicolas invented a story about hitting his elbow and getting sent to the Emergency Room by his Medical Agent.

"This morning the swelling was gone, so I took them off. All the same, I managed to do some composing. I've brought you the new smartfume."

"I didn't ask you to compose. I asked you to go to City Hall. Our customers can wait to sniff at something new, but we need to secure that contract. Think what we could do, Nico, with public inhalers mounted on every streetlight. An urban fragrance service to combat the stench of smog and the miasma from trash cans, to perfume every neighborhood, first in Rome, and then who knows where else?"

Pietro has had twenty magnificent fragrance stations installed in Piazza dei

Quiriti – booths, equipped with private alcoves, where people can retire in groups of two or, at the most, four at a time. Each station is furnished with an aroma diffuser, a nebulizer, and an incense burner. Customers breathe in the scent molecules, configured with atomic precision by the nebulizer, from one of eight small matter tanks attached to a stem. At regular intervals, which customers can adjust using the station's display, a nozzle opens, releasing a specific nebulized quantity of their chosen fragrance either into the booth or directly into their nostrils. The stations are easy to transport and, even when working round-the-clock, require only one refill per week, just like the old gas station tanks used to.

Nicolas fishes around in his pockets and stops dead on the threshold. Pietro does not fail to notice.

"What is it? Is something wrong?"

Inside, Nicolas is starting to panic, so he makes a show of looking around and admiring the new additions to the club.

"Not at all. You know, the stations have really turned out well."

"They certainly have. It's down to the space – the more you have, the better. Your commissioner friends were generous when they granted us that land-use permit. That's why it's important to me that you handle them with kid gloves."

Nicolas is struggling to remember where he left the phial. He's sure he didn't forget it at home. He's sure he brought it with him and that he was carrying it in his pants pocket.

"Come on. They just took down the drop cloths this morning. I'll show you how the renovations turned out."

The air inside Rendezvous is purified by three air exchangers, which stave off the threat of any unpleasant odor inside the club or any possible contamination being introduced from outside.

"With this lighting, the contrast between the various fragrances is heightened and, as a result, the customers' olfactory experience is enhanced. See what an effect it produces?"

In a corner, three different clouds are mixing and fusing together in the air, creating an impression like those old psychedelic seventies lamps filled with viscous liquids. Everywhere scents of frankincense and sandalwood envelop statues of different divinities set on altars in stone and bronze. In the inner courtyard, ringed by vases over three feet tall and ornamental plants, a papier-mâché elephant with dozens of candles around its feet stands before a furnace that is always lit, into which anyone can toss frozen or precooked, canned or microwavable food products, just to prove how little they care about food or material possessions. There are the clever ones who take advantage by burning expired or spoiled foods in it, but the other people don't know that. It's the gesture that counts.

Rendezvous is full of people, prey to an olfactory euphoria encouraged by conversations, confidences, kisses and other shows of affection. Ever since Fox Searchlight Pictures chose it to host a wrap party and Sky filmed a season of its sitcom *Sniff & Vape* there, every fashion designer, actor, dancer and soccer player who happens to be passing through has chosen it as their office-away-from-home and preferred recreation venue. Its clientèle consists largely of middle-class twenty- to forty-somethings who have made it their custom to frequent the club because they have, to various extents, abandoned traditional eating habits. They are tired of putting up with the old way of eating, sick of procuring food, preparing food and, most of all, disposing of the resulting waste. Whenever they're hungry, they simply turn to a nutraceutical from a taste distributor, from any Taste Station they can find, and they have no regrets.

Three women, lean as greyhounds, their shining faces 'touched-up' thanks to the surgical talents of some Artificial Visagetician, are perched on high stools in the light of columnar floor lamps. They are chatting, making as much noise as they can, and generally vying to be at the center of attention. At the station next to them, lying on a low couch upholstered in burgundy leather, a gay couple lie caressing, legs wrapped around each other, hands clasped together. 'He', the macho type, with a shaved head and black eyes, is weighing up 'her' fat wallet, which is lying on the coffee table. Meanwhile, 'she' – with a sweet gaze, thick wavy hair and speckled loafers – is eyeing the Jaguar key chain dangling from 'his' pocket.

At the bar, a pair of solitary customers are next to each other getting buzzed, each inhaling a black-coffee fragrance enhanced with hydroponically cultivated *Superskunk Oranje*.

"Give me a minute? I have to make a phone call."

"Fine, but then I want to try this new fragrance."

The ploy works and Nicolas steps away. At the end of the hall he shoves open a door that reads STAFF ONLY and hurries up to the second floor, taking long strides to avoid having to stop and say hello to Sudhir and Mirna in the fragrance designers' room. He reaches his office and shuts the door. What worries him most isn't the formula, which is saved in the memory of his nanomat and easy enough to reproduce, but the copyright for the smartfume, which is as yet unregistered. If someone files it before Rendezvous does, Pietro will accuse him of sloppiness and a lack of professionalism and make sure he loses sleep over it for a month.

Nicolas asks his smartphone to call the first number that pops into his head. "Hello, Police Emergency?"

He paces back and forth in front of his desk, curling the hairs of his beard around his fingers compulsively.

"I need to report a crime. A theft."

Someone knocks at the door, then sticks their head around the doorjamb without waiting for permission.

"Is everything all right, Nico?"

He places his finger over his lips to shush his mother and continues.

"Ah, the Authors and Publishers Association handles that? Fine. Is the form I need on their website?"

Olga, worried something serious is going on, comes closer. She is in charge of personnel and accounting at Rendezvous. Today she is wearing a modest knee-length beige dress made of nanotechnological fabric, cool in the summer and warm in the winter. With a simple brush of the fingertips it can become as smooth as silk or as soft as velvet.

"Thank you. I'll fill it out right away."

Today she is much shorter than him. She is wearing shoes with heels that can shrink or grow taller based on the height their wearer desires to be. At work, she chooses to cultivate a neutral image of herself.

She looks at him, blue eyes clouded with concern.

"My smartfume's been stolen, the new one. I'd just composed it."

"Who stole it?"

"That bastard of a rickshaw driver. I'm sure I had it in my pocket when I climbed in. Then I fell asleep – he kept on talking and talking – and a moment later it was gone."

Olga comforts him with a pat on the shoulder. "These things happen, Nico. Just don't tell your father you were sleeping. You know how he feels about certain things. He would take it badly and blame you."

"Yeah, I know. It's better if I tell him I got mugged, threatened with a knife."

"Are you sure?"

"No, it's just that I already told him yesterday that I hurt my elbow."

"And it wasn't true?"

"No." Nicolas walks Olga to the door. "It's a long story. I had something I had to do."

"And did you do it?"

"No, I got into an accident."

Nicolas is in a hurry. He has to send the form reporting the theft before the smartfume's formula can be copied or – and this is the worst-case scenario – extracted from the nasal mucosa of that Pulldogs pickpocket and allowed to circulate freely around Rome, scattering two weeks' worth of molecular permutations to the wind.

"Why didn't you tell me?"

"Because nothing happened. Edora was there. She helped me."

"She's a nice girl. Are you treating her right?"

Nicolas increases the pressure on his mother's arm and propels her out into the corridor.

"Of course. I wear kid gloves whenever I beat her."

"You joke about it, but it's plain to see that she cares about you."

"Speaking of caring, why don't you break the news to Dad? That way at lunch I'll just have to finish smoothing him over."

CHAPTER THIRTEEN

Aromas Come, Aromas Go

Pietro doesn't bother to knock on the door. He comes in while Nicolas is downloading the smartfume formula from his home nanomat.

"So, where's this fragrance I'm supposed to go crazy over?"

The lightness of Pietro's words can mean only one thing – that Olga's mediation has been a success.

"You've been telling me about it for days. Now, surprise me."

Nicolas launches composition. In just a few seconds, it's complete. He picks the phial up off of the tray, shakes it and sets it on the desk.

"There it is. You'll have to choose the name, as usual. I couldn't think of any good ones."

Nicolas was one of the first designers who had begun to specialize in volatile substances, the reason being that he does not much love other people, or rather, the way people's bodies smell. Every one of those biological organisms, regardless of whether man or woman, rich or poor, is, to his nose, an unwitting carrier of odors that are either insipid and wretched, or, at best, bland and uninspiring. If, given their hormonal constitution, men smell of dust, urine and sweat, then women stink of rancid fat and rotten fish. This is what Nicolas's nostrils perceive when he is close to other human beings, whose bodies are born without the protective and beneficial aura that a perfume can provide. Everywhere, whether in a gym or at the beach, at a café or in line at the cash register – and even where he is now, inside the offices of Rendezvous – Nicolas is assailed with an amalgam of awful, common, revolting and vulgar secretions.

Plants, the way he sees it, are far preferable. On the terrace of his apartment he has set up a pair of Chinese fan palms, each six and a half feet tall. The wall is masked by a philodendron that has climbed all the way up its trellis to the roof. In pots he has grown a *Eugenia caryophyllata* for its clove buds, a few yellow freesias flecked with orange and some ivory-colored tuberoses. On the landing outside his door, as a welcome, he has positioned a particularly pungent black hellebore. He has forbidden the Filipina who comes every Wednesday and Saturday morning from having any contact with *his* plants. He also has a little secret. He sows ailanthus seeds at random, tossing them into the air near subway stations, on sidewalks in front of gas stations and in parking lots. Nicolas always keeps a bag of them in the compartment underneath the seat of the Beast. Since he can't grow an ailanthus tree in his apartment, he's had to have the seeds sent to him from a nursery up the hill from the Baths of Caracalla. That tree is an implacable invader. It can put down roots in any crack and remain unobserved until its crown begins to peep out from between gratings, from inside fissures or any other tiny chink. If it isn't pulled out, roots and all, within a year an ailanthus can destroy a sidewalk or wreak havoc in a sewer system. Nicolas is very proud of his work. He finds the plants' appetite for destruction a splendid thing.

"How about the crime report? Did it get filed properly? Are we covered?"

"I sent it three hours ago. It's already been received and processed. Don't worry. The copyright should be safe."

"What do you mean, 'should be'? Didn't they send you a reply? Isn't there some procedure to be followed? Are we at the whims of whoever wants to live off what's rightfully ours?"

All it takes is one wrong word to ruin everything. There's always the risk that the smartfume was stolen on commission for large-scale reproduction. Up-and-coming perfume designers, trendy clubs competing with Rendezvous, dealers of psychedelic essences – all are potential suspects.

"The copyright *is* safe. These people don't mess around. They've assured me they're going to take immediate action. For stuff like this they work directly with the Anti-Nanoadulteration Squad."

Pietro appears mollified. Now that the immediate problem is out of the way, he can focus on pleasure.

"So, tell me, Nico. What's in this smartfume?"

"Methyl phenylacetate, but I removed the honey aftertaste, because it was too cloying. Then a few modified oxalic acid molecules from that Amazonian cacao you got last month. I bonded them with a carotenoid."

"Excellent. Simple yet effective."

"If you don't have time to evaluate it yourself, you could let the others 'sample' it first and see their reaction."

Mirna and Sudhir would have his back, even if he, the son of the owner, had moved up to the second floor. He hadn't asked for that promotion, nor did he consider it a step up in his career. If the choice were his, he'd still be downstairs with his fellow fragrance designers, weaving molecular bonds instead of human relations. The fact that Nicolas prefers working with raw materials from the inanimate world makes this all the more true. Often his orders contain molecules so rare and unusual that Mirna and Sudhir don't even know how to bond them. Once he has achieved an acceptable semi-finished product, Nicolas always invites them to drink it. 'Taste reinforces the sense of smell' is the phrase he uses to put an end to any uncertainty or reluctance on their parts. In his view, a fragrance designer who doesn't taste his creations – although Nicolas doesn't say this to anyone's face – is a sort of barbarian, or better yet, a coward who settles for using one sense only when he could use many at the same time.

"Perfect. I'll take it directly to the fragrance designers' room. Oh, and Nico, there's one more thing. There's a woman waiting in front of the club. She says she wants to speak to you. She's a strange one. Wait till you see her!"

"Who is she?"

"I don't know. I've never seen her before in my life. But you can't miss her. She's covered in tattoos and she's got a mohawk that's downright terrifying. She says she has to speak to you in person."

"A mohawk?"

Nicolas gets up and follows Pietro downstairs. When they get to the entrance to the club, Nicolas continues on outside, while his father, instead of heading into the laboratory, remains on the threshold to find out what he can about his son's business. As soon as he sees her, Nicolas can tell this is no ordinary woman. She's an Amazon, but with a rickshaw for a steed in place of a horse. An anti-smog mask covers her nose and mouth and her stance is aggressive, arms crossed over her chest, emphasizing well-developed muscles.

"Are you Nicolas Tomei?"

And yet, there's something familiar about her. His sense of smell triggers hazy mental associations and one particular odor – greasy, unpleasant, briny – is connected to something far in the past that Nicolas's mind rushes to dig up from memories of playgrounds in Villa Sciarra Park, of hiding places in the alleyways of Trastevere, of video-game tournaments with other middle-school troublemakers in the gambling halls around Piazza San Cosimato. His nostrils can recognize levels of toxic particulates better than if he had sensors implanted in his nasal mucosa. Nicolas was born and raised in the center of Rome, and even as a boy he could distinguish numerous substances – acids, poisons and resins – without getting them confused. On this stranger's body, beneath a layer made up of gasoline, ammonia, sulfur, the cement dust of construction sites and a hint of dioxin, Nicolas can smell a faint fragrance of grass, pollen, pine trees and packed earth.

"Yes, I am. Have we met?"

"Yes, we have, but you obviously don't remember."

Nicolas takes her by the arm to lead her away from the club's entrance, but the gesture is futile. She doesn't budge an inch. Olfactory stimuli continue to reach his nostrils: the sour smell of alcohol mixed with the stink of guano and the scent of an olive-oil-based skin cream.

"Would you mind if we didn't stand right here?"

His request is not so much due to his embarrassment at being seen with someone like her as his fear that Pietro will find him out in his lie – the logo on the woman's rickshaw, a sled dog, is the same as on the one belonging to the kid who stole the smartfume.

Slowly, they walk to the corner of the street and, once around it, she stops. She pulls something out of a pocket in the belt bag around her waist. "I believe this is yours."

In her outstretched hand is *his* phial of smartfume.

"Little Simon screwed up and he wanted me to tell you he's sorry. He's still a kid who sometimes thinks bad ideas are good ones."

Nicolas is distracted. He is studying the woman's features, her combative posture. He stares at her hands, dirty and neglected, a working woman's hands. With his head, however, he's elsewhere, trying to ferret out the possible origin of all those smells. Among all of his acquaintances, there are none who pull a rickshaw, none who wear a decorated anti-smog mask the way another woman might wear a silk scarf. Nor is there anyone covered in so many tattoos. Most importantly, there isn't anyone who smells like the *countryside*.

"How do you know it's mine?"

"I don't know if it's yours, but the label says Rendezvous. I asked Little Simon who he took the phial from, but he didn't know. So I asked him for a physical description, but it wasn't what I expected. Still, I knew that *you* might work here."

"You don't really mean to tell me that we've met?"

She peers back around the corner towards Piazza dei Quiriti to check on her rickshaw. "Does Via della Lungara ring a bell?"

It's been twenty years since Nicolas moved away from Via della Lungara, since Pietro bought the shop that today is home to Rendezvous and transferred his family to the Prati neighborhood. Nicolas stayed with his parents until he was twenty-eight and then, riding the crest of his success at work, he moved to Trastevere.

As suddenly as a light going on, it finally hits him. "Silvia Ruiz."

"Did you just throw out a guess or was that your usual bad luck?"

Bad luck. It's Nicolas's favorite phrase. It always has been. It's the phrase he used to use as an excuse every time that Silvia – Silvia Chili-Pepper, as everyone had called her back then – got him out of whatever latest batch of trouble he'd gotten himself into. Like that time at the Botanical Garden, when he'd been about to choke to death on a pine nut and she had slapped him on the back so hard he'd coughed it up. Or the time he'd gotten lost in the pine woods at Fabulus, the campground off of Via Cristoforo Colombo, and she'd been able to find him before the police arrived. Every time, Nicolas would blame it on his bad luck, and no one could ever make him admit to anything different.

"This time, though, I'd call it *good* luck. You've gotten me out of a very tight fix."

"Yeah, like that's a first."

There it is. Déjà vu. A rift in time. It's as though, after all these years, they are right back where they always were, with Nicolas in a mess, and Silvia pulling him out of it. They might be unrecognizable to one another now, after years of physical changes, but as children Nico and Silvia had been like brother and sister. Roberto, Silvia's real brother, was eleven years older than her. After his military service, just barely out of adolescence, he'd left for Australia and found a job in Sydney, working for a company that imported products from Italy. She hadn't seen her brother since – not even at Christmas, when down there it was summer vacation time – so Silvia had set Nico in his place.

"Can I offer you a scent?"

She stiffens, glancing down at her own powerful physique, her street attire, wild and unkempt. "Are you sure? I wouldn't want to scare away your customers."

"Are you kidding? After you've been so kind to me? Besides, I'm curious to know how you...."

"Got to be like this? I could ask you the same thing."

Nicolas takes the blow, lowering his gaze to the gut he's so ashamed of. If Silvia didn't recognize him at first, it's because he weighs around three hundred pounds. Nothing remains of his child's features on his adult face. His nose has become more prominent and curved. His eyes appear to have shrunk in comparison with his enormous round jowls. His wavy hair brushes his shoulders and his beard hides his double chin.

"Really. It would mean a lot to me to treat you to something. It doesn't cost me a thing. The club belongs to my father."

Silvia is astounded by his lack of tact. She would have preferred it if it 'cost' him something, given that she not only didn't resell his formula, but she made the effort to come all the way here to give him back his smartfume in person. In honor of a friendship that still has a place in her heart, she decides to overlook his thoughtlessness and accept his invitation.

"Fine, but I don't have much time. As soon as they call me, I'll have to run."

Escorted by Nicolas, Silvia makes her way through Rendezvous's curious and suddenly uncomfortable customers to sit at the bar. A gym-built blond, fed on steroids and dietary supplements, looks her up and down hungrily. The three simpering women whisper to each other at their reserved station as they enjoy the ritual of five o'clock fragrance-time. It's rare to see a rickshaw driver inside an aroma-bar, not because it's forbidden or too expensive, but because they themselves tend not to enjoy such 'sedentary' amusements. At most, couriers without vehicles of their own, who make deliveries and pickups on foot or on rollerblades, will often wait outside perfumeries, knowing that people forget all kinds of things there. Girls forget their handbags, hair clips, necklaces and earrings – all fresh from the nanomat – or else their expensive makeup, fake nails and false eyelash extensions, not to mention a veritable ton of smartphones. Guys leave behind their sunglasses, the keys to their motor-scooters, their shirts and jackets, wallets full of credit cards. When they call up the club, they can get their goods brought to them by a messenger who's already there, ready and waiting.

Rickshaw drivers are another race entirely. They gather in parks and empty lots, where they amuse themselves by competing at 'turnik', a Russian discipline that blends horizontal gymnastics with parkour, participating in exhibitions of extreme acrobatics, where the body represents the highest expression of sensory gratification. Other times you can see them leaping from rooftop to rooftop, dodging around cars at info-signals or engaging in complex choreographies that include climbing up onto or bouncing dangerously off vertical walls and falling from impossible heights. The most courageous among them play Urban Golf, a sport halfway between parkour and old-style golf, where 'holes' have to be made in as few shots as possible and the ball can never stop moving. If there are cars or pedestrians in the way or other obstacles blocking the 'course', it doesn't matter, so long as the endorphins keep flowing, guaranteeing that sense of happiness that is physical effort's reward.

Nicolas orders two 'effusors', little cups, each with a lid and a forked inhaler meant to be inserted into the nostrils. "How long has it been? It feels like a lifetime, doesn't it?"

"It might seem like a long time from the outside, but inside it hasn't been long at all."

"So, you're a rickshaw driver?"

"So, you make smartfumes?"

The Silvia and Nico who used to amuse themselves by throwing rocks at the Regina Coeli prison from the Janiculum Hill, calling out to inmates and pretending they were relatives, just to see what they'd do...that little girl and boy don't exist anymore. Their joy at seeing each other again can't even begin to scratch the layers of diffidence and feelings of estrangement that have built up between them, year after year.

Silvia stares straight ahead at her image in the mirror. Nicolas doesn't dare to look in any direction, for fear that Pietro, Olga, or someone else he knows will do something to embarrass him. Then, as the fragrance starts to fill his lungs, he begins to tell her about how he started at Rendezvous, of how he refused to leave to try his luck in Mumbai, Cape Town or Rio de Janeiro, where the best scents on the planet are composed. He tells her about how his father made him an excellent offer, ensuring him a very comfortable life right there in Rome. In exchange, he had taken Rendezvous and raised it to nationwide renown.

Despite the achievements he's describing, Silvia can hear a note of regret in his words. Unlike Nicolas, she has had an uncommon life.

"And how about you? Are your dad and mom still at Il Romoletto?"

Silvia places the effusor in her nostrils without taking her eyes off of the mirror. "My mother's well. She kept on running the restaurant after my father..." she takes a deep breath of the scent called Free Spirit, "...died."

"Died? What happened? When?"

"It was about ten years ago. A stupid robbery. And he was even more stupid." Silvia tightens her grip on the nozzle. Her nostrils flare rapidly, but not because of the fragrance. "He tried to defend the cash register, fought back like an idiot. He shot at one of the robbers and the robber shot back. Total damage: two men dead, one seriously injured, one child left without a father and two new widows, all for five hundred euros. Smart move, right?"

"Shit, I'm sorry. Were you there?"

"No, otherwise I'd have stopped him. Or maybe I'd be dead, too. I'd been gone for a couple of years already when it happened. I've never been able to stand that restaurant."

"Why? I remember we used to play hide-and-seek there. We'd hide under the tables or in the storeroom, or out under the ivy. One time I found you in the pizza oven."

She finally turns to face Nicolas. She shrugs. She's not in the mood to exchange confidences. He's not the person she expected to find, and she's not much like she used to be, either. A part of her is disappointed, another feels sorry for him. Nicolas's bulk is way beyond normal, pathological even, but the thing about him that has disappointed her the most is something less prosaic than his physical appearance.

Nicolas still doesn't know how to read Silvia's gestures, her movements, her silences and her half-formed answers, just like all those years ago. Underneath all the layers of pollutants, all he was able to do was identify her natural scent, that olfactory stamp that remains immutable throughout each of our lives, exclusive to each individual.

"You find me repulsive, don't you?" he asks.

Silvia goes back to staring at the mirror, then looks down at her dirty hands. "You know me, don't you? You know I can't lie."

Once, when he was twelve years old, Nicolas had been denied permission to go on a school field trip when Pietro dug in his heels, insisting it was pointless to miss a day of school just to visit the WWF nature reserve by Lake Bracciano. Pietro had left Nicolas at home to do his olfactory exercises but, as soon as he had left for work, Silvia had come over. The next day she had gotten a tongue lashing when, all innocence, she had told the teacher about it, and the teacher had immediately called Nicolas's mother.

"I swear, I've tried. I've been fighting against this body for eight years." He lands a smack on his flaccid chest, then another against his gut, which resonates beneath his white button-down shirt. "I've been to a dietician, an allergist, a psychologist. I've tried dissociated diets, vegan diets, the 'Zone' diet, high-protein diets, the glycemic-index diet, the blood-type diet. I've tried calorie-counting apps, metabolism apps, apps that calculate your dietary needs, but nothing's worked. I just can't."

"That's not what I was talking about."

Silvia looks around. Over her shoulder she can see Pietro making his rounds of club's stations, deferential and accommodating to every customer even though he's seething inside. She knows that falseness well, the hypocritical courtesy of someone who'll bleed you for your money with a well-honed smile. Pietro and her father lived across the hall from one another back on Via della Lungara. Indeed, the Ruiz and Tomei families had visited back and forth nearly every day for ten years. She feels like going straight over to Pietro and telling him how his neighbor died.

"Your father's an asshole, just like mine, no different. I saw him when you came out to meet me. He looks at you like you're still ten years old, maybe fifteen at best. Fuck, Nico. How can you live like this? Don't you see that every decision you make, every choice, every moment you don't rebel against him just makes things worse? You know your part in all this, and you just accept it?"

Pietro notices them and acknowledges them with a wave. Nicolas waves back.

Silvia ignores him.

"Now you listen to me, Silvia Chili-Pepper. You think you can just show up after twenty years and start stirring up trouble? My father may be bad tempered and a jerk. I'll admit he's got his hang-ups, but he helped me to become a *fragrance designer*. And he's going to sign this club over to me as soon as he retires. Everything you see here will be mine, and if you look around, you'll see it's worth quite a bit."

Silvia snorts a jet of fragrance out of her nostrils. "Do you think I give a shit about how much things are worth? Nice vases, classy lamps, a bunch of china and – oh, I almost forgot, a fake elephant covered in jewels. Congratulations. I want to live, not add things up, Nico. Assholes like your father, who are always pushing you to earn more money and count up how many things you have, I couldn't care less about."

She pauses to inhale, then keeps going. "Anyway, who could ever forget what your father's like? Do you remember that year we all went camping together? He was obsessed with the idea that the best cure for your bronchitis was to fill the air with 'balsamic vapors'. Have you forgotten about that? All day long that camper was like a eucalyptus gas chamber. At night he kept those three oil burners going, the ones with the candles, burning mint and rosemary oils, and that other one – what was it called? turpitude? – and camphor."

"*Turpentine.*"

"Right."

Nicolas remembers it well. His humiliation and discomfort had only been eased by the way he and Silvia had made light of them and of those ridiculous precautions which, in the end, had actually been the 'summer school' portion of Nicolas's rhinal education.

"You should have recognized it a long time ago. He doesn't care about anything or anyone. Well, at the least he doesn't care about anything but appearances. He cares how you dress, how much money he can squeeze out of you and how he can exploit you, but he doesn't give a damn about what's inside!"

A few heads turn in their direction and Silvia realizes she's gone too far.

"That's enough."

"Fine, I'm done anyway. It might not seem like it, but I didn't actually come here to lecture you. I only wanted to give you back your smartfume. I'm sorry."

That said, she gets up and turns to leave.

"No. Now you hold on a minute." He takes her by the arm and, this time, he uses enough strength to stop her. "I may be fat, I'll give you that, but I'm not an idiot, and I'm not going to feel insulted just because you have it in for everyone who's not like you. What, do you think that just because you have muscles and tattoos and spit on the ground and criticize everything that makes you a better person? Just because we Tomeis have money doesn't mean we're lazy or greedy or that we're cowards."

"It's not about money. It's about being domesticated. All the animals that live in the city are fatter, more promiscuous and more docile. Why should people be any different?"

"Is that so? And where the fuck do you live?"

Silvia turns to face Nico. He lets go of her arm. Something is about to snap inside her. Her chest is heaving like she's about to punch him in the face.

"I lived outside Rome for ten years, in Serra Spino, and for two years now I've been living on the Garbatella-Testaccio viaduct, the abandoned one. Me and the Pulldogs, we've made it into something new and now it's our home. Living up there isn't like living in the city, I guarantee it."

"Oh, no? And why's that?"

The way he remembers it, Silvia was always pigheaded, so why should she be any different now? Nonetheless, stubbornness aside, she does radiate a concealed energy. Close up, you can see it. Her profile emanates that magnetism, attractive and frightening, like that of a bird of prey about to launch itself on its quarry. Her head is lowered, her eyes fixed on him, her jaw clamped.

"Because the females of domesticated animals go into heat more often than wild ones. Horses, cats, cows, it makes no difference. Cats are lazy balls of lard, dogs put their ears back and make big eyes at you, and pigs get a whole lot fatter than boars do in the countryside. Need I go on?"

Nicolas takes a step back, unsure whether the reference to pigs was directed at him. "That may be true, but those animals were bred that way on purpose, to produce more, to fatten up more, to reproduce quickly. They have a better chance at survival if they do."

She lets her fist fall onto the bar instead of into Nico's face. Pietro is keeping an eye on them.

"Damn it, you just don't get it. I'm not talking about survival."

Pietro keeps watching them from behind, but he doesn't feel like getting involved. Silvia takes a deep breath and forces herself to calm down. "Sure, you're right. Pigs may live longer than wild boars, but do you call that living?"

She puts a hand on his shoulder, the way a friend would do. She squeezes lightly, as though about to share a confidence. "Look, quitting my job at the Post Office was the best thing I ever did. At least I was forced to take a look at the life I was leading, consider the alternatives and start really living. There are always alternatives. It's just that people either don't have the courage to imagine them or, if they do, to see them through to the end. To tell you the truth, it's a whole lot better to quit a job than to quit living. At age forty, there's still time for you to come back to life. I know it."

Nicolas doesn't get a chance to respond, because Silvia's smartphone starts to ring.

"Hey, Alan? What's wrong?"

She turns and walks away from the bar, but she's upset enough that Nicolas can still hear her perfectly.

"What? Your mother's been arrested?" She signals to Nicolas that she has to go. "Yeah, I'm on my way."

She turns back and reaches out, extending her hand to him.

"Thanks for the fragrance. It was very good." Her grip is strong and firm.

Nicolas has that dazed feeling you get just after leaving a job interview: *Thanks for your time, thanks for your interest. We'll be in touch.*

He can see her picking up speed before she's even left the club. He rises from his stool and follows her with his gaze. Silvia's legs, their muscles toned and well defined, propel her outside, behind the push bar of her rickshaw and then out into the flow of traffic, where she rapidly disappears from view.

Nicolas sits back down, his head slumped heavily between his shoulders. He stares at the effusor in front of him, then at Silvia's. He takes the phial of smartfume she brought back to him, sprays some on his wrist and sniffs at it. He forgot to ask her whether she or the kid, Little Simon, had inhaled the fragrance. The risk of counterfeiting remains, even if the phial is back in his possession. The presentiment he has, however, is much worse than that of a violated copyright. He has an overwhelming feeling that, along with the phial, Silvia has also restored to him the memory of the best part of his life. It's a painful feeling, because it's absolutely true.

Nicolas takes out his smartphone and calls his emergency number. "Hello, Edora?"

He's already putting on his jacket.

"Yeah, I'm calling about that evening you promised me. Change of plans. How about if I come over to your place?"

He pops his head into the fragrance designers' room and shoots them a look of inquiry. Sudhir and Mirna both give him a thumbs-up, signaling that the smartfume was a hit. He gestures to them that he's on his way out.

"What if I come over right now?"

He steps out of Rendezvous. Piazza Quiriti is full of passing rickshaws. Some are free, but Nicolas raises his arm and flags down a taxi.

CHAPTER FOURTEEN

A Rainbow in Black and White

"Oh, shit."

An acrid odor and a feeling of dampness. Nicolas is lying on his back. A sliver of pinkish light filters through the curtains and pierces his eyelids. Something serious has evaded the control of his nervous system. A splotch of urine has soiled his boxers and wet the sheet. He hastens through his usual fumbling journey towards the bathroom. He simply has to partially circumnavigate Edora's round bed, a distance of about seven feet, all the while avoiding tripping over poufs, then squeeze through the door without getting it mixed up with one of the room's three mirrors. After that, it's ten steps down the hall and five to the left. His bladder is aching and in his mouth he can taste a mixture of acid and bacteria that is almost nauseating. His blood sugar is below his daily level, but it is already late morning.

Once he's sitting on the toilet, a cramp in his lower abdomen doubles him over. He turns on his smartphone and lays his palm against the screen. It's a good time to measure his health and consult with his Medical Agent.

"Send the data home and give me today's verdict."

Real samples versus virtual samples. Actual data versus statistical data.

The result of the scan will help him to put together a breakfast that will prolong his life or, at the least, not shorten it. The best Indian doctors, whom he consulted on a health forum the previous month, prescribed him esoteric homeopathic compounds to be taken every two or, at the least, every three hours for a minimum of six months.

It is that sort of advice which, together with molecular prescriptions that can be printed up at the nanomat in the form of convenient pills and syrups, played a crucial role in the disappearance of the pharmaceutical industry and its outlets, the pharmacies, bringing an end to that oligopoly disguised as a free market, with its countless different labels but only one ultimate owner. In Nicolas's case, however, all they had helped him to do was lose six and a half miserable pounds.

"Yesterday's excesses have resulted in a life expectancy reduction of twelve hours."

Acerbic, as ever. The Medical Agent's sole interest when forming its assessments is to optimize his personal health – well, aside from relentlessly displaying nutraceutical banner ads and daily requests for him to participate in market surveys and trials.

In the end, why should a health app care about Nicolas's personal vices, monthly income or sexual encounters? It supplies simple unvarnished data, which is not meant to be viewed or interpreted in any mystical or philosophical light. It is for that reason that Nicolas tends to trust avatars more than he trusts human beings. Besides, he's the one who set it up that way.

The only drawback is that, if one day it gives him bad advice or, because of some random error, tells him he has six months instead of six weeks left to live, he would have no one to blame but himself. In any case, that's pretty much how things go with doctors, too. One time he'd gotten sick from something he ate at the Lariano Mushroom Festival. He'd woken up the next morning with terrible red spots on his forehead, neck, and chest. It turned out that the products that were supposed to have been not simply Italian but certified 'Protected Geographical Indication' had actually come from Romania. The doctor had told Nicolas he'd come down with food poisoning, asked him a few questions about his diet, and recommended a stomach pump and some cow's milk from his own farm.

Nicolas turns off his smartphone. One image from the previous evening refuses to leave his mind: Silvia.

He goes back to the bedroom. The time on the screen of the muted TV reads almost noon. Edora is stretched out on her side, her stomach protruding from the fitted sheet to spread along the edge of the bed.

They'd had sex twice when they got back home the previous night, and then a third time that morning after dawn. He'd liked the way she slapped him with her panties, an XL thong the color of a cantaloupe. Still, he can't help but think over and over about the muscular body of his old friend Silvia, her incredibly slender waist and her generous hips, shaped by exercise.

Without waking Edora, he pulls away the dirty top sheet. Luckily, there aren't any urine stains on the bottom one. He balls it up and goes off to shove it into the washing machine. He walks into the kitchen feeling better. Despite the sharp pains in his stomach and the death threats from his Medical Agent, he still has a feeling of postcoital relaxation, the kind that lasts for hours. He appreciates the generosity and softness of Edora's carbohydrate-shaped curves, those pliable handholds to caress, grab and squeeze, and yet, after the close-up view of the veins standing out in Silvia's neck, her sculpted calves and her taut, toned flesh beneath her tight T-shirt, he's beginning to nurse some serious doubts about his current sexual preferences.

The beverage compartment inside Edora's minibar is full of unsweetened tea, dealkalized water, low-calorie fizzy drinks, and two or three supplements lacking any clearer identification, composed by Edora herself based on who knows what sort of molecular formulas. On the food shelf is a neatly lined-up display of olive-flavored crackers, zero-calorie chocolate cake and a packet of six cherry croissants. Although a lot of people might turn their noses up at artificial flavors, or at least say they do, Edora is one of those who is so accustomed to the synthetic versions of certain foods that she prefers them to the originals.

For no particular reason, Nicolas finds himself staring at the furnishings. Edora is going through a veils phase. There are veils draping the bedroom and the same thing has happened in the kitchen. A ray of sunlight filters through the blinds, slipping between great cream-colored festoons that swell in the morning breeze until they brush against a Fashion World print of the Brooklyn Bridge at night that hangs on the wall behind them. The image calls to mind Silvia's viaduct. The fact that she's a childhood friend isn't a problem. Nico has discovered that he is attracted by *that* body, regardless of its associations.

He goes over to open the blinds and let some light in. Outside, far down Via Giulia, the rainbow at Ponte Sisto bridge is out of order, or rather, it's black and white. Maybe the marketing agency hasn't paid their freelancers, who have taken their revenge in the form of a sequence of grays.

Nicolas turns and looks at the stereo, a globe around which orbit six satellite speakers, identical to the one in Pietro's hobby room. He notes the modular eating utensils in resin and wood cellulose, the same ones that Olga gave him for his last birthday, perhaps even at Edora's suggestion. Then there are the chairs, the rug, the handles on the doors. Without even looking at them, he feels he knows them all, remembers them from some friend's place or another. It dawns on him that anyone could live as he does. Anyone could live in this apartment, or in any other, and possess the same exact things. If everything is interchangeable, then life becomes a copy of a copy of a copy of a life that could be anyone's. In a moment Nicolas sees himself condemned to a wretched future, the victim of customers obsessed with sniffing at the latest olfactory trend. He sees himself surrounded by middle-class trendsetters and product designers on contract with Kinea, their goal being to compose as many milligrams of product as they can from the raw material they're provided, or else by an army of pouting freelancers who all think they're the next Michelangelo or Leonardo da Vinci. It would be complicated to explain to Edora just how awful her kitchen is. Her sense of aesthetics aside, it makes him sad. While Olga might really like her and Pietro has never said anything against them seeing one another, what they have is certainly not any genuine sort of love.

Edora Frediano comes from an excellent family, crossbred from Mauro Frediano of the law firm Pessici & Tolzi and Sonia De Lollis, manager of the important Corso Trieste branch of Unicredit bank. What's more, she holds a degree from the Luiss Business School, she owns three apartments that a tourist agency runs for her as B&Bs, and she works as a copywriter at 69 ADS, a multimedia communications and advertising agency.

The moment Nicolas takes his smartphone off silent and checks his mail, he sees the response to the report he filed the day before with the Anti-Nanoadulteration Squad. His report has been received and, by the end of the day today, March 30, 2029, officers assigned to his case will have ensured that the perpetrator's nostrils have been subjected to 'remediation'.

"Shit, they didn't used to be this efficient."

A rumble of disquiet rapidly morphs into worry.

If Silvia has smelled the smartfume at all, even by accident, the Anti-Nanoadulteration Squad won't think twice about cleaning her out, too, and he would be sorry to see that happen, seeing how she went out of her way for him. Nicolas decides to warn her.

First, though, he turns on Edora's nanomat and selects the formula for milk. He extracts the lactose molecule and composites it. In the cupboard he finds some biscuits that don't set off the alarm on his Medical Agent and eats three of them. Five more get slipped into his pocket. Lastly, he links up with his own home nanomat and downloads a scent from his own personal archive of fragrances to leave as a gift for Edora. She kept her promise to help him have fun.

The previous evening she'd taken him to Nutrimancer, the number one spot in Rome for molecular taste explosions and compounds capable of activating the brain's own electric energy. The walls of the club are honeycombed with ornamented compartments, each containing an exquisite single serving of some delicacy assembled by one of the best food designers on the planet, all of whom reveal the secrets behind the conception and preparation of each recipe on their personal channels on the Yourfunfood network. In order to eat them, customers insert a hand into a compartment and fish out the silvery plastic tray it contains. A blinking green light signals each time a new tray arrives, sort of like on a film set.

After Nutrimancer, they had taken a short stroll up Via della Dataria to the Trevi Fountain to watch the spectacle of sounds and multicolored lights in the water. Once home, the moment they'd crossed the threshold, she'd slipped out of her heels and hiked her skirt high up above her formidable hips. She'd grabbed him by the hair and guided him down between her fleshy thighs. She'd wanted him to rip her nylons off with his teeth. After all, they were edible and mallow-blossom flavored.

Her feet in his mouth had tasted good.

When Edora gets up for breakfast during a break in her morning talk show, on the nanomat's tray she'll find a lily-and-jasmine-scented aromatic message spelling out, in airy characters, the words 'I'll see you soon. Kisses.'

It will put her in a good mood for the rest of the day.

CHAPTER FIFTEEN

The Pulldogs

At the entrance to the Garbatella-Testaccio viaduct hangs an enormous illegal billboard that reads:

NO MONEY NO CRY

It seems so old, without any wireless connection or built-in 3D graphics, that the marketing agents who mapped out the area for commercial purposes decided to consider it the property of the bridge's inhabitants, a derelict remnant of an unsophisticated and naive type of communication. The truth is that it's less than two years old, the result of modifications to a MoneyGram advertisement.

The previous evening, while at the Nutrimancer with Edora, Nicolas had managed to track down the formula he needed to re-composite the Beast's damaged fairing. He'd found it on the blog of some Lamborghini enthusiast living in Canada. They'd exchanged a couple of emails then moved into chat mode to talk about their personal mods. In the end, he'd traded the composition formula for the Belva's exhaust for the fairing from the Canadian model.

Nicolas parks his shiny Lambo, fresh from the nanomat, underneath the access ramp that runs up to the viaduct. He gazes at it with satisfaction, then activates its alarm system.

The product of the first formula he ever composed using Nanocad dangles around his neck, a toy whistle that he created during his first year at IED, the European Institute of Design. He had traded that one for a cup of coffee. Since then he has continued to accumulate formulas, saving the best ones for future trades. Everyone does the same thing, trading with everyone else so that there's never a need to create anything new. Rather, improvements are incremental, with each person adding a pinch of his or her own knowledge, experience and creativity. There are always those amidst the global server clouds who want or need a formula, or even a simple modification, and are ready to trade something for it. It might not even be a formula they really need. They do it for the mere pleasure of the exchange, to form a bond – two formulas, each granting the wish of the other.

The gate, etched with a blowtorch and decorated with phosphorescent graffiti, is open, so Nicolas enters without fear of committing a misdemeanor. From further above come cries and agitated shouts.

"Stop him!"

"Leave him alone!"

Then noises that resemble animal sounds – squawks, bleating and neighing. "Come on, catch him!"

A flock of green parrots rises into the sky. "I have nothing to do with it."

"Who the fuck do you think you are?"

Then comes the sound of rocks cracking against the ground.

Nicolas quickens his pace up the path of cracked asphalt. Lining the viaduct on either side of its four traffic lanes are numerous little houses made of bricks composited in some neighborhood PMC. Each dwelling has its own small vegetable garden that extends to the center of the median strip. Scattered about everywhere are dozens of rickshaws, some with chassis made of paper interwoven with carbon fiber, light but sturdy. Others are bizarre hybrids, which together appear to comprise every form of transport ever invented – Viking drakkars and Sicilian carts, bicycles built for two and streamlined skateboards. All are decorated, a superstitious precaution to ward off the risks and perils of journeying through the city.

"I haven't done anything! I didn't smell it, I swear!" The shouts repeat, the tone growing progressively more distressed.

"Hold him still! And the rest of you stay where you are or it'll be worse for you."

That which Nicolas is traversing with long strides is a micro-neighborhood suspended seventy feet in the air on the ramp that curves slightly then rises to span the Tiber upstream of Ponte Marconi bridge. It seems less a residential enclave than an encampment, through which snake bundles of cables and pipes, camouflaged by the growth of bushes, brambles and the sorts of trees you'd expect to find in a forest. Up here, the Pulldogs have been cultivating thousands of little plants that Nicolas recognizes – weeds and native grasses – perhaps chosen specifically because they don't require too much water. The roofs, meanwhile, are insulated with rags and fabric torn from used clothing from local flea markets. Over the last ten years, rain in Rome has become scarce. When it gets any at all, it's concentrated in brief downpours.

The strangest part of the scene is the bridge's pylons. While they originally must have been intended solely as support elements, they now appear to have been modified. Above the deck level, where they rise to form towers, enormous holes have been cut to hold solar panels, making them look like ancient lamps.

Nicolas comes to an open space and before him is the scene he'd feared to find. In the middle of a gathering of people, some of whom are dressed in workman's coveralls, three officers from the Anti-Nanoadulteration Squad have immobilized the boy whose name, Nicolas recalls, is Little Simon, the self-confessed – well, to Silvia, at least – criminal. He is on his knees, two officers holding him still, one to each arm, while the third is holding an 'aneffusor' to his face. It is a latex nose-glove which, once it has adhered to the nostrils, will extend two slender prongs into the boy's nasal ducts, where they will identify and, if they find them, cauterize the neurons in his olfactory epithelium where the memory of the smartfume has been recorded.

"Now we'll see if you were telling the truth."

Nicolas runs up just as Little Simon's nose lights up like a blue light bulb. "No! No!"

The third officer holds the boy in a headlock. The prongs have crept up to his olfactory bulb and are examining the cilia. Little bluish sparks flash here and there inside his nostrils. Anywhere they find the formula, they are burning the olfactory endings, causing temporary anosmia.

"Wait. A moment, please."